STUDYING FOR CHEMISTRY

LARRY FRANKLIN LITTLE
DeKalb College

BRENDA D. SMITH, Series Editor
Georgia State University

 HarperCollins*CollegePublishers*

Acquisitions Editor: Ellen Schatz
Cover Designer: Ruttle Graphics, Inc./D. Setser
Electronic Production Manager: Angel Gonzalez Jr.
Publishing Services: Ruttle Graphics, Inc.
Electronic Page Makeup: Ruttle Graphics, Inc.
Printer and Binder: R. R. Donnelley & Sons Company
Cover Printer: The Lehigh Press, Inc.

Studying for Chemistry, First Edition

Library of Congress Cataloging-in-Publication Data
Little, Larry Franklin, 1946–
        Studying for chemistry / Larry Franklin Little.
        p.   cm.
        Includes index.
        ISBN 0-06-500651-8   (pbk.)
        1. Chemistry—Study and teaching.   I. Title
    QD40.L585 1995
540'.71—dc20   94-37734
                                                          CIP

95 96 97 9 8 7 6 5 4 3 2 1

Dedication

*This book is dedicated to the memory of my parents—
Gillis Dowling Little and Ida Mae Holley Little—and to
my brothers and sisters: Kenneth Little, Carl Little,
Sharon Little Riggs, and Vickie Little Harris.*

TO THE STUDENT

This book is written for you. The purpose of this book is to provide you with a resource to use along with your textbook in studying and learning chemistry. The primary goal of this textbook is twofold:

1. to provide you with models for learning chemistry concepts, and

2. to help you develop problem solving strategies to apply to chemical quantitative concepts.

Throughout this textbook you will find suggestions on how to apply different types of thinking skills for learning and understanding challenging chemical concepts.

Many beginning chemistry students have difficulty analyzing and solving chemistry problems. One feature of this textbook is the development of solution pathways for solving many kinds of chemistry problems. You will find detailed descriptions of the thinking process, as well as plans for setting up and solving many of the quantitative problems associated with beginning chemistry.

Another feature of this textbook is Applying the Concept: Exercises. These exercises are found after each topic or concept discussed in the textbook. The questions allow you to practice or apply the skill or concept presented in the topic, thereby evaluating or monitoring your comprehension of the topic. The answers to all Exercises are found in the Appendix so that you can check your progress.

Learning chemistry is an active process. Hopefully, this textbook will help you in developing learning strategies that will allow you to become more active in reading your text, note-taking in lecture, and studying for lecture tests and exams. I wish you a very positive experience in your study of chemistry!

CONTENTS

CHAPTER 5 THE HEART OF CHEMISTRY: Organization of Matter   59

CHAPTER 6 THE PROCESS OF CHEMISTRY: The Power of the Balanced Equation   81

CHAPTER 7 WET CHEMISTRY: Solute, Solvent, and Solutions   103

CHAPTER 1

WHAT IS CHEMISTRY ALL ABOUT?

GETTING FOCUSED

- *As you begin your study of chemistry, what are your biggest concerns or apprehensions about being academically successful in your chemistry class?*

- *How do you think your approach to studying chemistry will be different from the way you study and prepare for other classes?*

- *Are there other thinking skills besides memorization that you will need to use and apply to learning chemical concepts and solving chemical problems?*

- *What are higher-order thinking skills (HOTS) and how do you apply these thinking skills to learning chemical concepts and solving chemical problems?*

CHEMISTRY AS CONTENT AND PROCESS

It is absolutely amazing and wonderful how the science of chemistry impacts the quality of our everyday lives. Chemists are responsible for the development of plastics, artificial sweeteners, Freon used in air conditioners to keep us cool, synthetic rubber, and synthetic fibers such as rayon and polyester. Synthetic fertilizers and insecticides developed in chemistry laboratories greatly increase crop yields, thus increasing food supplies. Chemists working in the field of biochemistry and pharmacology have made important contributions in the development of drugs that save the lives of millions of children and adults every year. Biochemists are in the forefront of recombinant DNA engineering that has the potential to increase the world's food supply, alter the devastation of genetically inherited diseases such as muscular dystrophy and cystic fibrosis, and to help put violent criminals behind bars. This list is only a tiny fraction of the many contributions chemists have given to the world—and the future holds many more secrets waiting for another generation of research chemists to unravel.

Even if you are not planning to be a research chemist, a working knowledge of chemistry is still important in the study of many other fields, such as biology, nursing and health sciences, nutrition, and medicine. Whether you are taking a course in chemistry as a chemistry major or whether you are taking it to meet the requirements for another field of study does not matter; the most important thing is to get the most from your first chemistry course. Getting the most from your first chemistry course means that you leave the classroom and laboratory with chemical knowledge, skills, and an appreciation for chemistry as a science as well as a process of discovery.

You probably already have a pretty good idea of what the study of chemistry is all about. If you have been reading in your textbook, you probably came across the old standby definition of chemistry: the study of matter and the changes it undergoes. Don't be fooled by this overly simplistic definition of chemistry. The study of **chemistry** involves learning and using terminology, concepts, theories, models, and problem solving to understand chemical phenomena. The study of chemistry also involves scientific inquiry. The chemistry laboratory will allow you to be involved in the process of chemistry; that is, investigating the physical and chemical changes that matter undergoes. Hopefully, in the laboratory the standard textbook definition of chemistry will take on new meaning as you actually see some of the changes that matter undergoes.

GOING BEYOND MEMORIZING

A research study was undertaken several years ago to determine how students used their textbook in learning a particular subject. The researchers found that the vast majority of students believed the most important thing to study in each chapter were definitions. Consequently, most of their efforts went into memorizing terms. Many chemistry students also share this viewpoint and are disappointed and upset when quizzes and exams prepared by their professors require them to go beyond vocabulary. Instead of solely defining terms, chemistry students are frequently asked to explain physical and chemical phenomena, predict outcomes, apply knowledge in problem-solving situations, draw valid conclusions from data sets, and defend answers to both qualitative and quantitative problems.

Let's be honest! The simple truth is that you, as well as thousands of other students sitting in introductory chemistry classes, *cannot* memorize your way to "success" in learning chemistry. To be successful in chemistry you are going to have to learn to "think" like a chemist. Granted, you have to know facts in order to think in any discipline. An important part of learning chemistry is becoming familiar with the terms and concepts that comprise the language of chemistry. Yet there are many qualitative (descriptive) and quantitative (numerical) concepts in chemistry that require mental strategies other than "recall" or knowledge level learning in order to fully understand the chemistry involved. Chemistry is an "active science" and requires "active mental participation" in the classroom, in reading the text, in working homework problems, and in the laboratory.

Cognitive psychologists (scientists who study how people learn) call these active mental strategies used in hypothesizing, predicting, controlling variables, defending arguments, drawing valid conclusions, and problem solving "higher-order thinking skills" because they require the learner to go beyond memorizing. When using higher-order thinking skills, a learner takes information and applies it in many different ways to get new information or to find answers to problems. Knowing about higher-order thinking skills and how to apply them appropriately to solving chemical problems is the key to success in learning chemistry.

TYPES OF HIGHER-ORDER THINKING SKILLS

Benjamin Bloom and his assistants conducted an analysis of items used by teachers to determine students' achievement in subject matter. Using the results of the study, the researchers grouped the learning tasks into six levels of difficulty, thereby creating a hierarchy of tasks similar to the arrangement used in biology to classify plants and animals. Instead of using Kingdom, Phylum, Class, Family, Genus, and Species, Benjamin Bloom and his associates identified knowledge, comprehension, application, analysis, synthesis, and evaluation as the different levels of thinking required to answer various types of questions found on classroom tests and exams. In this hierarchy, knowledge questions are considered the least difficult to answer, and evaluation questions are considered the most difficult. These learning tasks or skills are referred to as the **higher-order thinking skills** or HOTS. Table 1.1 describes the six levels of learning tasks and gives a description and examples of each. It is important that you take the time to understand the different levels of higher-order thinking skills so that you may make use of the appropriate skill when it is needed for understanding chemical concepts and problems you will encounter during your study of chemistry.

You might be a little confused about the types of higher-order thinking skills listed in Table 1.1. When can you tell if a topic or concept requires a higher-order thinking skill for understanding the topic or concept? How do you specifically apply the thinking skill for comprehension? These are very important and legitimate questions to ask. The answer is of course you have to be taught this skill. You need to be able to analyze each topic in your textbook, your lecture notes, your laboratory textbook, and the printed handouts that your professor might use as lecture supplements and rate them as to the type of thinking skill required. Then you need to apply the appropriate thinking skill to your reading, studying, writing, and problem-solving activities. Each of the following chapters in this study skill book will give you examples of when and how to apply specific higher-order thinking skills in learning and comprehending chemical topics and concepts.

APPLYING THE CONCEPT: EXERCISE 1.1

Suppose you are studying for a chemistry test, or completing a lecture, or a laboratory assignment. What kind of thinking skill would you use or apply to fully understand or complete the following tasks? Some of the tasks require the use of several higher-order thinking skills in order to complete the task. Use Table 1.1 in Chapter 1 as a resource. Pay particular attention to the second and third columns in this table.

1. Defining the term *osmosis.*
2. Performing a titration in the lab to determine the percentage of acetic acid in a vinegar sample.

TABLE 1.1   HIGHER-ORDER THINKING SKILLS

LEVEL OF THINKING SKILL	WHAT THE STUDENT KNOWS OR DOES USING THE SKILL	ACTION VERBS FOR SPECIFIC OUTCOMES
Knowledge or recall (Giving it back)	Knows terms Knows rules Knows specific facts Knows classifications and categories Knows criteria Knows methods and procedures Knows principles and generalizations Knows theories and structure	Defines Names States Identifies Describes Distinguishes
Comprehension (Translating it)	Translates communications Interprets relationships Extrapolates from given data	Interprets, converts, explains, predicts, generalizes, infers
Application (Trying it on problems)	Applies principles	Uses, solves, constructs, prepares, demonstrates
Analysis (Taking it apart)	Analyzes organizations and relationships	Discriminates, outlines, diagrams, differentiates, infers, explains
Synthesis (Creating a new work)	Produces new arrangement	Designs, organizes, rearranges, compiles, modifies, creates
Evaluation (Making a professional judgment)	Judges on basis of external criteria Judges on basis of evidence	Appraises, compares, contrasts, discriminates, criticizes, detects

3. Writing a report on the impact of heavy metallic ions on the environment.

4. Reading a description of a chemical experiment and determining which variable described is the:

   independent variable.

   dependent variable.

   controlling variable(s).

5. Completing a set of assigned problems on concentrations of solutions after attending a lecture where your instructor explains and

demonstrates the techniques of problem solving relating to concentration of solutions.

6. Your instructor calls on you to go to the chalkboard and write in detail, the solution to a chemistry problem. You are then asked to give an oral explanation to the class on how to solve the problem.

7. On a quiz, you are asked to list four criteria for determining whether or not a chemical reaction has occurred.

8. For a laboratory assignment, you are asked to design and outline a plan to separate a solution containing several cations and to identify each cation in the solution.

9. For an essay question on a lecture exam, you are asked to discuss the difference between *covalent* and *ionic bonding.*

10. You are assigned to read an article on causes of air pollution and to critique the article on the basis of:

    - the accuracy of the chemistry discussed in the article.
    - the feasibility of the author's suggestions for reducing air pollution.
    - your own recommendations for ways to clean up the atmosphere.

TERMS TO KNOW

chemistry
higher-order thinking skills

CHAPTER 2

TAKING CONTROL OF LEARNING CHEMISTRY

GETTING FOCUSED

- *How can you use your textbook more efficiently to extract and retain chemical terms, concepts, and problem-solving skills?*

- *What is the purpose of taking lecture notes? Is there a better way to organize your lecture notes to help you in preparing for tests and exams?*

- *How can you improve your thinking strategies for solving chemical problems?*

MAKING THE MOST OF YOUR TEXTBOOK

Just like many other beginning chemistry students, you will depend heavily on your textbook to provide you with basic information needed to build a knowledge base in chemistry. True, you will attend lectures, listen, participate in class, and take notes, but obviously you cannot take your instructor home or back to the dorm to refresh your memory or fill in the "gaps" in your lecture notes. Consequently, it will be to your benefit to learn how you use your textbook as an effective learning tool. There is a great deal of information in your chemistry textbook. Your job is to recognize the different types of prose found in science textbooks and to apply the appropriate thinking skill in order to retain and recall the major concepts presented. This is no easy task! Research has shown that the major reason science students have trouble reading scientific textbooks is that they lack a framework to guide them in selecting what is important and in organizing information so that it can be recalled and used.

In *Thought and Knowledge: An Introduction to Critical Thinking,* Diane Halpern, a cognitive psychologist at California State University at San Bernardino, cites research that supports the fact that students can be taught to use their textbooks more efficiently. Dr. Halpern describes a method known as **structure training**—a method by which students are taught strategies for recognizing, comprehending, and organizing the type of expository prose that is found in science textbooks.

What are the types of expository prose found in science textbooks? The best way to find the answer to this question is to open your chemistry textbook to the beginning of a chapter and find examples of the types of prose that are discussed below.

Macrostructure

Every chapter in your textbook is broken down into units of prose describing a main idea or concept. These units or main ideas constitute the **macrostructure** of the chapter. The macrostructures are usually printed in large, bold colored text and are easy to recognize. The macrostructures correspond to the chapter outline as listed in the table of contents of your textbook. Take a moment and look at the Chapter 3 outline in the table of contents of your chemistry textbook. Now, go to Chapter 3 and locate each macrostructure. You will notice that each macrostructure presents a new topic, idea, or concept and is followed by one or more pages of expository prose (text) explaining the topic, idea, or concept.

APPLYING THE CONCEPT: EXERCISE 2.1

List four topics found in Chapter 3 of your textbook.

It should be obvious that if you understand and can recall the main ideas presented in each topic or macrostructure of the chapter, you have successfully learned the chapter. Unfortunately, many students think they understand the material as they read, only to find that the knowledge is not in their heads when the textbook is closed. How can you be assured you understand fully each topic in the chapter before you start a new section or macrostructure? First, take responsibility for monitoring your own comprehension of the material in each unit. Stop after each paragraph, passage, or section, close your textbook, and write or recite aloud the main idea or ideas in your own words as if lecturing to your class. You should also list or recite supporting and disconfirming evidence related to each idea. This is the real test to determine if you truly understand the ideas or concepts presented in the macrostructure and if you are ready to go on to the next topic in the chapter. It is extremely important for you to write down or recite aloud your responses. When you write, or recite, you are thinking, and when you are thinking, you are organizing information into memory that makes recall and retrieval of that information more assessable.

What kind of thinking skill are you using in restating the main ideas(s) of each macrostructure in your own words for the chapter you are studying? Refer to the list of higher-order thinking skills presented in Table 1.1 on page 5. As you can see from the table, you will use the comprehension level of higher-order thinking skills. The general objective is to interpret relationships and translate the meaning of the topics and ideas presented into your own working knowledge of the study topic.

Depending on the nature of the material presented, there may be instances in the macrostructure where you will use other higher-order thinking skills besides comprehension. If the macrostructure you are reading deals with a quantitative topic illustrating methods of solving chemical problems, you will more than likely use the application level of higher-order thinking skill. Here the skill is to study carefully the worked-out examples of the chemical problem presented in the macrostructure of the textbook. In beginning chemistry, almost all problems can be solved in one or more steps by using conversion factors, ratio and proportion, or a simple algebraic equation.

The steps required to solve the problem are called the **solution pathway.** In a quantitative macrostructure, the main idea or topic is concerned with developing a solution pathway for solving similar problems. Once you understand the solution pathway for the chemical problem presented, you apply the thinking process (solution pathway) to similar problems to see if you fully understand the mathematical steps involved. When you are monitoring your comprehension after studying a quantitative macrostructure, find similar problems at the end of the chapter and apply the solution pathway to as many problems as possible.

APPLYING THE CONCEPT: EXERCISE 2.2

Locate the chapter in your textbook on measurement. Find the macrostructure that explains dimensional analysis or the "factor label" method of working measurement problems. List the steps one applies when working problems using dimensional analysis or the "factor label" method of solving measurement problems.

If you still have difficulty understanding the mathematical steps (solution pathway) needed for successful solution to the problems, then it is time to seek help. There are many resources available. You just have to ask! Your instructor or teaching assistant is required to keep office hours for the specific purpose of helping students. Contrary to popular belief, your instructor wants to help you be successful in your study of chemistry. After all, your instructor is the expert, and expert help is the best kind! Before you make an appointment to see your instructor, make sure you have invested time and mental energy in seeking solutions to the problems presented in the macrostructure of the textbook. Go to the study session prepared to show your instructor your attempts at problem solving. Be prepared to discuss specific areas in the solution pathway that are giving you difficulty with the chemistry problems presented in the topic. It also helps to have a set of carefully thought-out questions to ask during the help session. Your instructor will appreciate the fact that you have made an initial mental effort to learn the assigned material and have invested time in thinking through the solution to chemistry problems. The end result is that you will walk out of the help session with answers to specific problems and your instructor will now know you as a student who is focused, conscientious, and dedicated to learning and understanding chemistry.

Other resources include your classmates, a study group, and the math or computer lab. Ask your instructor if computer chemistry tutorials are available for student use. Many chemistry tutorial software packages do an excellent job of presenting the mathematics of chemistry. Unlike your instructor or study mate, the computer is very patient and forgiving! If it takes you several times to get the answer correct—no problem! The computer just keeps on looping to the topic and allowing you to review and practice more problems until you finally get the hang of it!

Sequences

In chemistry textbooks, **sequences structures** are those in which events are discussed in a specific order. Sequences must be learned in order. A typical example of a sequence structure in chemistry textbooks is a passage describing how to write a Lewis structure for a molecule. A Lewis structure is a special way of writing the structural formula of a molecule to show how the electrons are bonded. Look in the appendix of your textbook and find the page numbers that discuss Lewis structures. You will note that at the very beginning of the discussion your textbook will list the steps involved in writing a Lewis structure for a molecule. The steps are actually an algorithm (a problem-solving plan). If you follow the steps in order, you will always end up with the right answer. There are many other instances in your textbook where information will be presented in sequential form. The key to understanding and retrieving information successfully is to realize that a series of steps must be followed in order to successfully solve a problem, draw a structure, balance an equation, list a series of events, or carry out a titration. Learn to recognize sequential structures in your text so that you may apply the appropriate learning strategy for comprehension.

APPLYING THE CONCEPT: EXERCISE 2.3

On what page of your textbook is the Lewis structure explained?

Causal Relationships

In chemistry textbooks, there are many examples of passages that describe events of causal relationships. In a **causal relationship,** event A causes event B to occur. For example, adding colorless liquid A to colorless liquid B causes a bright yellow precipitate (solid) to settle out when liquid A and liquid B are mixed. Of course, the key concept here is understanding the chemistry involved to make the event happen.

Since chemistry is an experimental science, the study of causal relationships is very important. Many of the discoveries, laws, and theories of science are the result of determining or finding out causal relationships between two or more variables. Finding the relationship between variables is what the **scientific method** is all about. In most chemical experiments, there exist three types of variables. The manipulating or **independent variable** is the variable that the researcher deliberately changes during the course of the experiment. The responding or **dependent variable** is the variable that changes in response to a change in the independent variable. Other variables that might affect the outcome of the experiment and must be kept constant are called **controlling variables.** Let's use Boyle's Law as an example. The independent variable (pressure) is deliberately changed during an experiment and its effect on the volume (dependent variable) is measured. The results of such an experiment show that, if you increase the pressure on gas, its volume will decrease. If you decrease the pressure, the volume of the gas will increase. In chemistry, this type of relationship between two variables is called an **inverse relationship.** In this example, the data collected during the experiment constitutes the evidence needed to establish the causal relationship between the pressure and the volume of a gas. Many causal relationships you will encounter in your chemistry textbook will involve measured variables. As you read and study these relationships, try to identify the independent and dependent variables and determine the relationship between the two variables.

The successful study strategy to apply for understanding causal relationship structures in your chemistry textbook is to identify the antecedent event and the consequent event and find evidence in the textbook to support that the antecedent event caused the consequent event to happen. What type of higher-order thinking skill will you need to use to organize causal relationships into memory for future retrieval and use? Again, if you will refer to Table 1.1 on page 5, you will see that the most appropriate higher-order thinking skill to apply is analysis. When you apply the analysis higher-order thinking skill, you identify the antecedent and consequent event and think through and establish the relationship between two variables.

APPLYING THE CONCEPT: EXERCISE 2.4

In the chapter in your textbook in which the gases are discussed, find the pages that discuss Gay-Lussac's Law. In Gay-Lussac's Law, which variable is the independent variable, which is the dependent variable, and which is the controlling variable?

Enumeration

Chemistry texts often contain passages that list or describe facts about a given topic. This type of structure used in scientific texts constitutes an unordered list of events. The key element here is unordered. You do not have to memorize the lists in order. Some examples of **enumeration structures** found in chemistry textbooks include the (1) statements comprising Dalton's Atomic Theory, (2) types of chemical bonds, and (3) types of chemical reactions. When you encounter enumerations passages in your textbook, you should be able to identify and list the information discussed in the text. A quick reference to Table 1.1 shows that the type of higher-order thinking skill to apply for understanding enumeration passages in your textbook is the Recall or Knowledge level of thinking. In enumerations passages, you are simply recalling or restating specific facts in list form.

Comparisons

Comparison structures are used in chemistry textbooks to point out similarities and differences within a topic. Examples of comparisons include comparisons of types of chemical bonding, comparisons of acids and bases, comparisons of mixtures and compounds, and comparisons of suspensions, colloids, and solutions. A good strategy for learning comparison passages in your textbook is to make two or more lists that describe the similarities and differences in the topics being compared.

What type of thinking skill will you need to use to understand comparison passages? Suppose you were given an unknown liquid chemical and all the reagents (chemical solutions) and apparatus you need to find out if the unknown liquid was an acid or a base. In the course of your investigation, you discover that blue litmus paper turned red when dipped into the unknown liquid; the unknown liquid felt "slippery to the touch;" and when a probe from a pH meter was placed into the unknown solution, the meter registered 11.7. If you had previously made a list comparing the physical and chemical properties of acids and bases, you would know immediately that the unknown liquid was a base. When you make a judgment based on evidence either stated or collected involving comparing and contrasting, you are using the Evaluation level of higher-order thinking skills. You might want to remember that comparison questions are a favorite type of short answer or discussion questions used by chemistry instructors on exams!

Response

Response structures are questions posed in the textbook and then answered in passages that follow. The response structure may be simple in that it can be answered in a single statement, or it may be more complex and require a sequence list, enumeration list, causal relationship, or comparison response to provide a complete answer. Suppose the textbook question begins, "What are the most important characteristics of the ionic bond?" The prose that follows the question in the textbook would answer the question by giving the most important characteristics of the ionic bond either in a list or paragraph form. Although different response structures require different methods of providing answers, your task for understanding the material presented in the response structures is to identify the question and the answer(s) identified in the textbook. As you might suspect, the thinking skill required for understanding material presented in a response format could range from Recall level to Evaluation level depending upon the complexity of the question(s) being presented.

Definitions

Many students complain that the terminology used in chemistry textbooks is "alien" or foreign to them. One reason for this reaction is that many beginning chemistry students try to memorize definitions of terms isolated from other terms or from the context of the topic or concept that the definitions describe. Words or terms by themselves are of little informational value. Only when the terms or definitions are used to develop ideas or concepts does the true meaning of words become clear. When you come across **definition structures,** always keep in mind that a particular term is important in order to understand a particular chemical topic, idea, or concept. Make a list of definitions as you encounter them in your study. Write a short concise meaning for the term in your own words and identify the idea or concept to which it is related. This will allow the word or term and its associated meaning to be linked mentally to a broader idea or concept. This kind of studying and learning leads to a "connected" knowledge base and facilitates the retrieval and use of the term from memory when needed.

■

PERSONAL LEARNING QUESTIONS

Do you think your approach to reading and studying in your chemistry textbook will be different after reading this section? Do you understand that different types of expository prose found in your chemistry textbook require that you apply different thinking strategies for comprehension?

■

TAKING CONTROL OF CHEMISTRY IN CLASS

Going to class and listening to your instructor go over the main points presented in your textbook and illustrating problem-solving strategies is an important part in learning and comprehending chemistry. Yet, sometimes it is very difficult to listen attentively during class and write down information in a format that makes sense when you study your notes later. What is the purpose of lecture or class notes? How should you organize your notes as the class proceeds? How should you use your lecture notes along with your textbook in studying and preparing for tests and exams? These are very important questions. Once you know the answer to these questions, you will have a powerful tool to help you in your pursuit for academic success in chemistry.

What Is the Purpose of Taking Lecture Notes?

First of all, lecture or class notes are a mirror reflecting terms, concepts, ideas, and problems that your instructor deems important. Your lecture notes contain your instructor's interpretation of material presented in the textbook or from other printed sources such as scholarly journals that represent the discipline of chemistry. Most instructors follow very

closely the topics presented in the textbook, but select the most important parts of the chapter to discuss in class. More than likely, these are
the topics, concepts, and problems that will appear on your lecture
tests and exams. Pay particular attention if your instructor goes over information not contained in your textbook. Write down the source of
the information stated by your instructor in case you need to go to the
library and look up the article or material for further study. If your instructor does not give the reference for the out-of-textbook material in
lecture, don't be afraid to ask for the source. Asking for the source only
implies that you are motivated and interested in the lecture topic, and
professors like these qualities in their students.

How Should You Organize Your Lecture Notes?

One of the most useful strategies for organizing lecture notes for understanding and comprehending material presented during class comes
from a study from Cornell University—*Learning How to Learn*. In applying this note-taking strategy, students are asked to take a ruler and,
using a pen or pencil, divide the pages of their lecture notebook into
two halves. At the beginning of the lecture, write the main topic to be
discussed during the class at the top of the left side of the page. If the
topic changes during the lecture, write the name of the new topic on
the left side of the lecture page. During the lecture or discussion, write
down the most important information presented by your instructor on
the right side of the lecture page. If you cannot remember all the information presented, or need more information leave a blank space on
the right side of your notebook and use your textbook to fill in the
"gaps" of missing information later. At the end of the lecture, all the information you have recorded will be on the right side of the page and
the topic names will appear on the left side of the page. Figure 2.1 is a
sample page of a student's lecture notebook using this strategy.

What about the left side of the lecture notebook? Why waste all that
space and paper, you may ask? The space and paper will not be
wasted! The reason for leaving the left side blank except for the topic
name is to enable you to convert your lecture notes on the right side of
the notebook into a series of questions on the left side of the notebook
as you study and review your notes. The end result is that you have
identified a large percentage of the questions, and most importantly,
the answers to most of the questions that your instructor could possibly

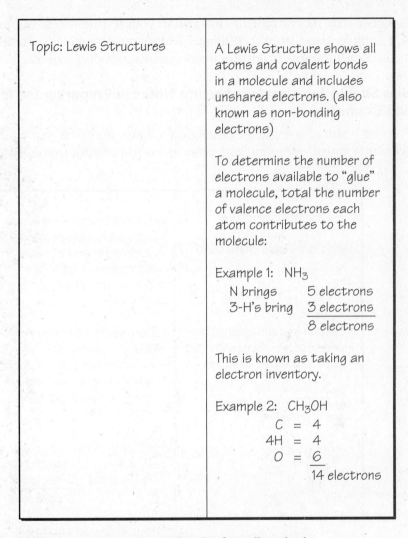

| Topic: Lewis Structures | A Lewis Structure shows all atoms and covalent bonds in a molecule and includes unshared electrons. (also known as non-bonding electrons) |

To determine the number of electrons available to "glue" a molecule, total the number of valence electrons each atom contributes to the molecule:

Example 1: NH_3
    N brings   5 electrons
    3-H's bring   3 electrons
                         8 electrons

This is known as taking an electron inventory.

Example 2: CH_3OH
                C = 4
            4H = 4
              O = 6
                       14 electrons

Figure 2.1 A sample page using the Cornell method

ask you on lecture tests and exams. Isn't that a wonderful thought! It is also a good idea to rate each of your lecture questions as to the level of higher-order thinking skill required to answer each question. You can rate your questions by referring to Table 1.1 on page 5, which describes the types of higher-order thinking skills used in finding answers to questions or solutions to problems. Once you have identified the difficulty level of the question, you can apply the appropriate thinking skill necessary for learning the material in your lecture notes. Figure 2.2 is a sample page from a student's lecture notebook illustrating the method of turning your lecture notes into a series of questions.

How Should You Use Your Lecture Notes in Preparing for Tests and Exams?

The end result of using the Cornell note-taking strategy is that you end up with a set of questions and answers corresponding to the material

Topic: Lewis Structure What is a Lewis Structure?	A Lewis Structure shows all atoms and covalent bonds in a molecule and includes unshared electrons. (also known as non-bonding electrons)
How can you determine the number of electrons used to "glue" a molecule together?	To determine the number of electrons available to "glue" a molecule, total the number of valence electrons each atom contributes to the molecule: Example 1: NH_3 　N brings　　5 electrons 　3-H's bring　3 electrons 　　　　　　　8 electrons This is known as taking an electron inventory.
What is the electron inventory for CH_3OH?	Example 2: CH_3OH 　　C = 4 　4H = 4 　　O = 6 　　14 electrons

Figure 2.2 Turning the lecture notes into a series of questions

covered in lectures and in the textbook. To monitor for comprehension as you study your lecture notes, cover the right side of your lecture notes that contain the answers to your questions. Now, ask each question out loud and give an answer to the question verbally, or even better, give a written response to the question. Can you answer the question? If not, uncover the right side of your notes and review the material you wrote down. You might need to find the corresponding material in your textbook to see if your notes provide the breadth and depth needed to answer fully the question posed on the left side of your lecture notes. If you continue this process until you can provide correct responses to all of your lecture questions, then you should be adequately prepared for any upcoming test or exam covering the material contained in your lecture notes.

Will this method of organizing your notes work for you? You will never know if you don't give it a try! What if you try this method of note taking, and the questions asked by your instructor on your test or exam are not even remotely similar to your lecture questions? If this happens, perhaps you need to discuss fairness in testing with your instructor. If your instructor is honest and fair in his or her testing methods, then the questions appearing on lecture exams should reflect the knowledge and skills presented in class and from textbook reading assignments. An honest and open discussion devoid of hostility and irrational behavior on your part is the best way to resolve any concerns you have about fairness in testing with your teacher.

■

Personal Learning Questions

What method do you presently use in taking lecture notes? Do you think the Cornell note-taking method might work for you?

■

TAKING CONTROL OF CHEMISTRY HOMEWORK

Homework is an important part of learning chemistry because it gives you the opportunity to apply the chemical knowledge and skills presented in lecture and class discussions and from reading assignments from the textbook. Most of your chemistry homework will consist of working out the solutions to problems found at the end of each chapter or in problem sets handed out by your instructor. Consequently, you need to develop good problem-solving skills and apply those skills to completing your homework assignments.

What is needed is a "model" to apply to solving chemistry problems assigned for homework. We previously defined a solution pathway as a series of steps needed to solve a chemistry problem. Let's use the idea of a solution pathway to develop our chemistry problem-solving model. Our model will be based on the diagram below:

Known or Given → Connections → Goal or Unknown

In this problem-solving model, the given comprises all the variables listed in the problem and their numerical values. The goal or unknown is the answer to the problem, usually represented by a number and a unit, such as 5.0 grams or 3.4 moles. The connections are all mathematical information needed to get from the given to the unknown. Most of time the connection(s) will be a conversion factor, ratio and proportion, or a simple algebraic equation. The connections you need and their order comprise the solution pathway needed to get from the known to the unknown. In simple problems, the solution pathway may consist of only one step. But most chemistry problems will require two or more steps in the pathway to arrive at the correct answer.

The best way to understand how to use the problem-solving model is to follow an example. Consider the chemistry problem below:

A chemistry student needs 140 millimeters of aluminum wire for an experiment. The student has only a ruler calibrated in inches. How many inches of aluminum wire does the student need to cut to provide the needed 140 millimeters of wire?

1. Write down the problem-solving model just as it is shown below.

    Known or Given → Connections → Goal or Unknown

2. Write down all given variables by name and numerical values under the Known or Given column and identify the unknown variable or goal state under the Goal or Unknown column.

    Given or Known → Connections → Goal or Unknown
    140 mm Al   inches of Al wire

3. Identify the connection(s) needed to define the solution pathway to obtain the desired answer.

Known or Given → Connections →   Goal or Unknown

140 mm Al $\dfrac{2.54\ cm}{1\ in}$ or $\dfrac{1\ in}{2.54\ cm}$ inches of Al wire

$\dfrac{10\ mm}{1\ cm}$ or $\dfrac{1\ cm}{10\ mm}$

4. Set up the problem in a series of steps using the identified connection(s) and check to see if the problem will yield the desired unit.

$$140\ \cancel{mm}\ Al \times \dfrac{1\ \cancel{cm}}{10\ \cancel{mm}} \times \dfrac{1\ in}{2.54\ \cancel{cm}} = inches\ (desired\ unit)$$

5. Carry out the mathematical operations identified in the solution pathway and express the numerical part of the answer to the correct number of significant digits. Make sure your answer contains the right unit as well as the correct number of numerical digits.

$$140\ mm\ Al \times \dfrac{1\ cm}{10\ mm} \times \dfrac{1\ in}{2.54\ cm} = 5.5\ inches\ Al$$

Another model used frequently in solving chemistry problems is the **factor-label method.** This method is sometimes referred to as **dimensional analysis.** This model is very useful because it establishes a relationship between different units expressing the same physical dimension. In the factor-label method of problem solving, units are treated mathematically like algebraic quantities. A problem is set up so that unwanted units cancel out of the problem and the desired unit is kept throughout the calculation. The factor label method makes use of **conversion factors.** A conversion factor is a fraction that expresses the relationship between equivalent quantities. Some examples of conversion factors are:

$$\dfrac{12\ in}{1\ ft} \qquad \dfrac{1000\ mm}{1\ m} \qquad \dfrac{2.205\ lb}{1\ kg} \qquad \dfrac{2.54\ cm}{1\ in}$$

When solving problems using the factor-label method, apply the following steps in developing a solution pathway and solving the problem:

• Read the problem carefully for comprehension and determine what the desired goal is or what you are being asked to do.

• List all the given variables and their units given in the problem.

- Identify the unknown variable and the correct unit for the unknown.

- Identify conversion factors needed to convert the known to the unknown.

- Set up the given variable and the conversion factors in such a way that all unwanted units cancel out and the desired unit remains in the solution.

- Carry out the mathematics and check your answer for the correct number of significant digits.

The following example will illustrate the use of the above steps in solving problems using the factor label method.

Convert 7.55 pounds to kilograms. One kilogram is equivalent to 2.205 pounds.

1. In this problem we are being asked to change pounds into kilograms.

2. Given: 7.44 lbs

3. Unknown: equivalent weight in kilograms

4. $\dfrac{2.205 \text{ lbs}}{1 \text{ kg}}$ or $\dfrac{1 \text{ kg}}{2.205 \text{ lbs}}$

5. $7.55 \text{ lbs} \times \dfrac{1 \text{ kg}}{2.205 \text{ lbs}}$

6. $7.55 \text{ lbs} \times \dfrac{1 \text{ kg}}{2.205 \text{ lbs}} = 3.42 \text{ kg}$

Using the problem-solving model presented will help you to break the problem down and analyze the components of the problem. It will also help in developing the correct solution pathway that leads to the correct numerical answer and the unit called for in the problem. If you will take the time to set up solutions to problems in this manner, it will provide you with a framework to "think" your way through the solution process. When you "think" your way to solving problems rather than memorizing a series of steps, you develop skills that will allow you to solve similar types of problems as well as develop solution pathways to new and novel problems.

■

PERSONAL LEARNING QUESTIONS
What is your current approach to solving quantitative problems? Do you see the advantage of planning your solution before you actually begin the problem-solving process?

■

APPLYING THE CONCEPT: EXERCISE 2.5

Using either the Given-Connection-Unknown model or the factor-label model, carry out the following conversions:
1. Convert 2.4 gallons to liters. (4 qts = 1 gallon, 1 liter = 1.06 quarts)
2. Convert 7.20 feet to meters. (12 inches = 1 ft., 2.54 cm = 1 inch, 100 cm = 1 meter)
3. Convert 388 grams to pounds. (454 grams = 1 pound)

TERMS TO KNOW

causal relationship
comparison structure
controlling variable
conversion factors
definition structure
dependent variable
dimensional analysis
enumeration structure
factor-label method

independent variable
inverse relationship
macrostructure
response structure
scientific method
sequences structure
solution pathway
structure training

CHAPTER 3

IMPORTANT CONCEPTS IN CHEMISTRY

GETTING FOCUSED

- *List five chemistry topics and briefly describe what you think is the most important idea or concept about each topic.*

- *Chemists are constantly taking measurements as they work in the laboratory. Name at least five types of measurements and give an appropriate unit for each measurement.*

- *Look at the Table of Contents in your chemistry textbook. Which chapters do you feel that you have at least some prior knowledge in which to build new knowledge about the chapter topic?*

- *Look at the Table of Contents in your chemistry textbook. Which chapters do you feel that you have little or no prior knowledge in which to build new knowledge about the chapter topic?*

Every subject you study in your college career has a core set of concepts that defines the knowledge base of the subject. Chemistry is no
exception. As you attend lectures, study in your chemistry textbook,
and attend the chemistry laboratory, these concepts will gradually become incorporated into your working knowledge of chemistry. Let's
take a brief look at some of the most important concepts you will
more than likely encounter in your study of chemistry. It is important to
realize that understanding the central ideas behind these concepts is
critical for succeeding academically in the classroom as well as in the
laboratory.

MEASUREMENT

First of all, chemists do not use the **English system** of measurement.
They use the **Metric system.** For example, the meter is the base unit
for measuring length, the gram is the base unit for measuring mass, and
the liter is the base unit for measuring liquid volume. In the metric system, one describes how much length, mass, volume, and other physical quantities by using prefixes in front of base units such as a kilometer, a centigram, or a milliliter. The most common metric prefixes and
their numerical values are listed on the chart below.

COMMON METRIC PREFIXES

MEGA	KILO	DECI	CENTI	MILLI	MICRO
1,000,000	1000	0.1	0.01	0.001	0.000001

There are other metric prefixes used to express extremely large and
small numbers that your instructor may assign for you to memorize.
Your textbook should have a complete listing of metric prefixes and
their numerical values.

There is no way of getting around the fact that you must become
comfortable working with metric units and prefixes. Invest the mental

energy upfront in learning the Metric system thoroughly. It will pay off many times in future assignments, laboratory reports, and problem sets that require answers in metric units.

APPLYING THE CONCEPT: EXERCISE 3.1

Which of the following contains the greatest amount of mass?
1. 400 g
2. .540 kg
3. 1000 mg

Chemists and chemistry professors are also very particular about the way numerical answers to problems involving measured units such as length, mass, density, and temperature are expressed. Your chemistry instructor as well as your laboratory instructor will expect you to express your answers to problems and measurements taken in the laboratory to the proper number of significant digits. **Significant digits** are all the units in a measurement that are known for certain, plus the first digit you have to estimate. The bottom line about significant digits is that there is a limit to the number of digits you can record when you take a measurement. The number of digits you can record in a measurement depends upon the precision of the instrument used to take the measurement. Beginning chemistry students often get into trouble when they add, subtract, multiply, or divide measured numbers using their calculators. Don't make the mistake of recording all the digits your calculator displays! It is inaccurate as well as dishonest to record an answer to a problem involving measured quantities to the thousandths place when you used a ruler that will only measure to the tenths place. Learn what significant digits are and why they are important in measurement. Almost every chemistry textbook has an appendix in the back explaining significant digits and giving many examples of how they are used in measurement, especially in the laboratory. Take the time at the very beginning of the course to review the appendix in your textbook on significant digits. Work all the practice problems involving significant digits until you become proficient and comfortable with this important part of measurement.

APPLYING THE CONCEPT: EXERCISE 3.2

How many significant digits are contained in the following measured numbers?
1. 103,650
2. 0.00456
3. 104.00

THE ATOMIC THEORY

The **atomic theory** is one of the simplest theories you will encounter in your study of chemistry. This theory forms the framework of our knowledge of the atom. The atomic theory was formulated by John Dalton (1767–1844), an English school teacher who based this theory on chemical knowledge that existed at the time. Dalton used the research of other scientists to formulate the assumptions of the atomic theory. The basic assumptions of the atomic theory are:

- Elements are composed of tiny particles called atoms.

- The atoms that make up a given element are identical.

- The atoms that make up different elements have different properties such as number of subatomic particles and masses.

- Atoms retain their identity during chemical reactions; they are not changed. Atoms are not created or destroyed during ordinary chemical reactions.

- Compounds result when two or more atoms of different elements chemically unite.

- For every chemical compound the kind and number of atoms is constant.

Don't be alarmed if the atomic theory statements in your textbook are not phrased exactly like the above statements; the wording may be different but the ideas should be the same. Study each statement carefully and then relate all statements together as a unified whole. The atomic theory explains many observations of matter, and with a few minor modifications, Dalton's atomic theory is still accurate.

APPLYING THE CONCEPT: EXERCISE 3.3

When Dalton formulated the atomic theory, isotopes of elements were not known to exist. Look up the term *isotope* in your textbook. How would you rewrite the second statement of Dalton's atomic theory to account for the existence of isotopes?

ORGANIZATION OF MATTER

Since chemistry is the study of matter, how matter is organized is extremely important. This topic usually appears at the very beginning of your textbook. Key concepts of this topic include elements, compounds, and mixtures. Physical and chemical properties of matter are also important components of this concept, as well as physical and chemical changes. Energy changes that accompany chemical reactions are usually introduced under this topic. The key concept of energy changes is heat energy and how it is measured. Pay particular attention to the units of calories and joules, how they are related, and how they are used to express the heat content of a chemical system.

ATOMIC STRUCTURE

The fundamental particles of the atom are protons, electrons, and neutrons. As you study this concept, try to visualize these particles in terms of location, size, mass, and electrical charge. The idea of atomic weights of the atoms of the elements is extremely important. You need to understand the process by which chemists measure atomic weights and to recognize the importance of atomic weights as being relative weights. Electronic configuration, which describes the way electrons are distributed around the nucleus of the atom, is a critical concept to understand because it plays an important role in explaining the concept of chemical bonding and chemical reactivity of elements.

THE PERIODIC TABLE

The **Periodic Table** is one of the most powerful tools available to the chemist and the beginning chemistry student. The power of the table is that it brings instant access to important information about the elements. Most textbooks contain a copy of the Periodic Table on the front or back inside cover as well as a copy in the chapter covering the Periodic Table. Knowing the position of an element in the Periodic Table along with an understanding of the numbers located above and below the symbol of the element in the table allows you to predict physical attributes of that element as well as its chemical behavior as it interacts with other elements.

The information in the Periodic Table is not meant to be memorized. The table is a reference source that you will make use of throughout your study of chemistry. It is very important for you to study the historical development of the Periodic Table to understand its arrangement of the elements in columns and rows. Make sure you take the time to understand the organization of the table and the significance of the quantitative information (i.e., the numbers above and below the symbol of the elements) given for each element in the table. A good understanding of how to use the Periodic Table will serve you well in your quest for academic success in chemistry.

CHEMICAL BONDING

The concept of bonding describes the forces that hold chemical particles together in compounds. Chemical bonds are continuously being broken and formed when elements and compounds enter into chemical reactions. Chemical bonds always involve electron rearrangement between atoms in close proximity to each other. The way the electrons interact between two atoms defines the type of bonding responsible for binding the atoms together as a chemical unit. **Ionic bonds** result when electrons are completely transferred from metallic atoms to nonmetallic atoms. **Covalent bonds** result when two nonmetallic atoms share electrons. Polar covalent bonds result when two nonmetallic atoms share electrons unequally. Only a select few of the electrons in an atom participate in forming chemical bonds. This is why a working knowledge of electronic configuration of the atoms of the elements is extremely important. It turns out that only the electrons in the outermost shells of atoms participate in bond formation.

APPLYING THE CONCEPT: EXERCISE 3.4

Look at the formulas for the following compounds. Which compounds are covalent? Which compounds are ionic?

1. NaBr
2. CO_2
3. HCL
4. K_2O
5. NH_3

THE MOLE CONCEPT

Chemists carry out chemical reactions for the particular purpose of making chemical products. These products may be pharmaceutical drugs, fertilizers, agrichemicals, or numerous other products. Chemical reactions take place at the molecular or ionic level. Yet, when chemists are carrying out reactions in the laboratory, they mass out chemical reactants in grams, kilograms, or metric tons.

How do chemists know how much of chemical A is needed to react completely with chemical B to make a given amount of product C? The answer to this very important question requires an understanding of two important concepts: the balanced equation and the mole concept. Research has proven that the atomic weight of an element or the formula weight of a compound expressed in grams contains a definite number of particles known as the **mole.** The number of particles in a mole is 6.02×10^{23} chemical units and is known as **Avogadro's Number.** One mole of atoms, electrons, molecules, or ions always contains 6.02×10^{23} particles. The reason that this number is so incredibly big is that atoms, ions, and molecules are so incredibly small. Therefore, a mole in chemistry defines a definite mass of a chemical substance that contains a given or fixed number of particles of that substance just as a dozen defines a unit amount of 12 and a gross defines a unit amount of 144.

APPLYING THE CONCEPT: EXERCISE 3.5

Which of the following contains Avogadro's number of particles?
1. 13 grams of NH_3
2. 18 grams of H_2O
3. 20 grams of HCl

The beauty of the mole concept is that it allows chemists to weight a given mass of elements or compounds containing a known number of particles in such a way that when the chemical reaction is finished, almost all of the reactants are completely converted into a desired chemical product. In the chemical industry where making chemical products for profit is the name of the game, it is critical to know the exact masses of chemical reactants to make the process cost efficient.

This stage is where the balanced equation comes into the picture! The numbers, or coefficients in front of the elements and compounds participating in the chemical reaction (i.e., $2N_2$, $3H_2$, $2NH_3$) and the co-efficients in front of the products formed, give the ratio of combination of the reactants and the ratio of formation of the products.

The following equation shows the commercial formation of ammonia known as the **Haber Process:**

$$N_2 \text{ (g)} + 3H_2 \text{ (g)} \rightarrow 2NH_3 \text{ (g)}$$

Industrial chemists using the mole concept know that 1 mole or 28 grams of nitrogen gas always combines with 3 moles or 6 grams of hydrogen gas to form 2 moles or 34 grams of ammonia under ideal conditions. If you don't know where the numbers 28, 6, and 34 came from, remember the definition of the mole and refer to the atomic weights for nitrogen and hydrogen on the Periodic Table of your textbook.

APPLYING THE CONCEPT: EXERCISE 3.6

Consider the following reaction:

$$2 N_2 \text{ (g)} + 5 O_2 \text{ (g)} \rightarrow 2 N_2O_5 \text{ (g)}$$

What do the coefficients 2, 5, and 2 represent?

FORMULA WRITING AND BALANCING EQUATIONS

Being able to write the formula correctly for chemical compounds is a crucial skill. Formula writing is important because you have to be able to write formulas for compounds to balance chemical equations properly. You have to balance an equation correctly to work stoichiometric problems successfully. **Stoichiometry** deals with the mathematics of the balanced equation, such as determining how much of a reactant you need or how much of a product can be formed in a given chemical equation. The following is an example of a stoichiometric problem:

How many grams of ammonia can be made if 20 grams of hydrogen gas reacts with an excess of nitrogen gas?

Do you see that if you cannot write the formulas correctly for ammonia, hydrogen gas, and nitrogen gas you cannot balance the equation properly? If you cannot balance the equation properly, you cannot determine the correct set of coefficients (the numbers that go in front of each reactant and product). Since the coefficients determine the ratio of hydrogen gas to nitrogen gas and the amount of ammonia that is formed, your final answer to the problem will be incorrect if the coefficients are incorrect. Therefore, be persistent and patient as you study the section of your textbook dealing with writing formulas for compounds and balancing equations. The key to success here is practice, practice, and more practice!

KINETIC THEORY AND GASES

The **kinetic theory** is important in explaining the physical state of matter; that is, whether a substance is a solid, liquid, or gas under normal conditions. The theory relates the two concepts of heat and motion in determining the physical state of matter. The kinetic theory is particularly important in explaining the properties of gases. In your study of gases, you will use five laws to solve mathematical problems dealing with gases. These laws are listed in Table 3.1.

TABLE 3.1   GAS LAWS

NAME	DESCRIPTION
Boyle's Law	Shows mathematically the relationship between the volume and pressure of a gas at constant temperature.
Charles' Law	Shows mathematically the relationship between the volume and temperature of a gas at a constant pressure.
Gay-Lussac's Law	Shows mathematically the relationship between the pressure and temperature of a gas at constant volume.
Combined Gas Law	Combines Boyle's, Charles', and Gay-Lussac's Law into one unified mathematical expression.
Ideal Gas Law	Shows mathematically the relationship between the pressure, volume, temperature, and number of particles (moles) of a gas.

As you can see, the study of gases requires a lot of mathematical problem solving using the above gas laws. As you work problems dealing with gases, read the problem carefully, make a list of the variables named in the problem, and identify the unknown (the answer you seek). A review of the variables list will give you a good idea of the type of gas law problem the question represents. Identifying the type of problem is the first step to success in problem solving dealing with gases.

SOLUTION CHEMISTRY

Many chemical reactions are carried out in water and other liquid solvents. In the human body, all of the chemical reactions that make up our metabolism are carried out in an aqueous (water) environment. When you place a solute (the substance being dissolved) into a solvent water, it greatly affects the physical and chemical properties of water and results in the formation of a solution. The chemistry of solutions is a very important concept in clinical chemistry. It is important to understand the different methods of expressing the strengths of solution such as percent concentration, molarity, molality, and parts per million (ppm). Other important topics include how water dissolves solutes and colligative properties of solutions. **Colligative properties** of a solution are those properties that depend entirely on the number of dissolved particles in the solution. Important colligative properties include freezing point depression, boiling point elevation, vapor pressure lowering, and osmolarity. If you are majoring in the health professions, pay particular attention to the concept of osmolarity. Fluids that are injected into patients must have similar concentrations of dissolved particles as internal fluids such as blood and tissue fluids. If they do not have similar concentrations, water will flood across cell membranes, causing cells to rupture, or water will be drawn out of cell membranes, causing the cells to collapse.

ACIDS, BASES, AND SALTS

These three types of chemical substances are collectively known as **electrolytes** because, when dissolved in water, they form ions and conduct an electrical current. Acids and bases have a unique set of chemical properties that you need to know to understand the various types of chemical reactions in which they participate. Some of the more important types of chemical reactions are ionization of acids in water, dissociation of bases and salts in water, neutralization, and reaction with other chemical substances. The strength of acid and base solutions is another important concept because it relates to ionization constants, pH, and pOH of acid/base solutions. Your calculator will be a useful tool in solving acid/base problems. You need to become familiar with the log key and exponent key on your calculator as they will be used frequently in solving mathematical problems dealing with acids and bases.

■

PERSONAL LEARNING QUESTIONS
Now that you have had a brief introduction to important topics in chemistry, which topics do feel "comfortable" about in terms of learning and understanding the topic? Which topics do you feel will be the most challenging in terms of learning and understanding?

■

TERMS TO KNOW

atomic theory
Avogadro's Number
colligative properties
covalent bond
electrolytes
English system
Haber process

ionic bond
kinetic theory
Metric system
mole
Periodic Table
significant digits
stoichiometry

CHAPTER 4

MEASUREMENT AND NUMBERS IN CHEMISTRY

GETTING FOCUSED

- *What is the most appropriate method of expressing extremely large and small measured numbers in chemistry?*

- *What are significant digits and why are they so important in recording measured numbers?*

- *How do we measure length, mass, volume, and temperature in the laboratory?*

- *How are measured variables used to determine physical properties of chemical substances?*

- *What problem-solving strategies work best for solving measurement problems?*

- *How can you use your calculator more efficiently to solve measurement problems?*

TAKING INVENTORY

Making, recording, and manipulating measurements mathematically is an important part of "doing chemistry." Chemists frequently measure mass, volume, temperature, boiling points, freezing points, wavelengths of light emitted by excited electrons, and atomic masses of atoms. Measurements such as these are used to determine the physical and chemical characteristics of elements and compounds, as well as to identify unknown chemical substances. Consequently, making and interpreting measurements is an important skill used throughout the study of chemistry.

Do you have a good understanding of the concept of measurement as it is used in chemistry? Do you know how to identify the appropriate measuring instrument and how to record properly measurements taken in the laboratory? Do you know how to mathematically manipulate measured numbers and the unit of measurement they represent? A good way to approach any new topic in the study of chemistry is to take a learning inventory about that topic. A set of learning exercises related to different aspects of measurement in chemistry is listed on page 39. Don't be disturbed if you cannot answer all of the questions. The purpose of the learning inventory exercise is to identify the "gaps" in your knowledge and skill level about a particular chemical topic. Stop now and complete Learning Inventory 1 before continuing.

CLEARING UP MISCONCEPTIONS

After completing Learning Inventory 1, check your answers in the Appendix. How well did you do? It is important that you clear up any misconceptions you might have about measured numbers and measurements. After studying the rest of this chapter, you will have the opportunity to "rethink" any misconceptions you may have had about measurement.

As you study this chapter, it might be a good idea to have your textbook open to the chapter on measurement. Read each macrostructure (heading) in this textbook first. When you finish reading about a topic, find the related topic in your chemistry textbook. After carefully reading the related topic in your chemistry textbook study all sample exercises. If sample problems are presented on the related topic in your textbook, try to work out the solutions. Chemistry students often find it helpful to read about a topic from two or more viewpoints.

Learning Inventory 1

Read each of the following statements. If you agree with the statement, check the "True" box. If you disagree with the statement check, the "False" box.

STATEMENT	TRUE	FALSE
1. If you use the right instrument and measure carefully, your measurement will be free of errors.		
2. In the metric system, mass is measured in grams, length is measured in meters, and volume is measured in liters.		
3. All measurements have a limit to the number of digits you can record to describe that measurement.		
4. All measurements consist of two parts: a number and a unit.		
5. Accuracy and precision in measurement really mean the same thing.		
6. When you add, subtract, divide, and multiply measured numbers using a calculator, the correct answer is the answer displayed on the answer window of the calculator.		
7. Scientific notation is a method of expressing extremely large and small measured numbers.		
8. Measured numbers and their units are not treated the same in mathematical calculations.		
9. For all practical purposes, mass and weight have the same meaning.		
10. Exact numbers have no uncertainty in measurement, whereas measured numbers always have some uncertainty.		

EXPRESSING SMALL AND LARGE MEASURED QUANTITIES IN CHEMISTRY

Chemists and chemistry students frequently use extremely large and small measured and exact numbers in chemical calculations. For example, you already know that one mole of any chemical unit contains 602 000 000 000 000 000 000 000 particles of that chemical unit. The mass of typical lead atom is 0.0000000000000000000035 grams. Can you imagine having to write these numbers out frequently and trying to keep up with all those zeros? Fortunately, there is a more efficient way of expressing large and small numbers. **Exponential notation** is based on powers of 10. Below are some examples of exponential notation:

$$1,000,000 = 10^6$$
$$10,000 = 10^4$$
$$1000 = 10^3$$
$$10 = 10^1$$
$$1 = 10^0$$
$$0.000001 = 10^{-6}$$
$$0.0001 = 10^{-4}$$
$$.01 = 10^{-2}$$

In these examples, 10 is the base and the superscripts are exponents indicating the number of times the base must be multiplied or repeated as a factor. If we multiply 10 times itself six times, we get the number one million. In exponential notation, the number one million is written as 10^6. The exponent "6" is known as the power of the base. We read "10^6" as 10 raised to the sixth power.

Not all numbers encountered in measurement in chemistry are even powers of ten. When numbers are not in even powers of 10, we used a form of exponential notation known as **scientific notation.** When numbers are placed into scientific notation, they have the following form:

$$M \times 10^N$$

In this formula, M represents the coefficient (any number greater than one but less than ten) and N represents the exponent or power of ten. Large numbers have positive exponents and small numbers have negative exponents. When we place large numbers in scientific notation, we move the decimal point to the left and stop just before the last digit. For example, light travels at a speed of 186,000 miles per second. In scientific notation, 186,000 is written as 1.86×10^5. In this notation, 1.86 is the coefficient and 5 is the exponent or power of ten. How did we end up with 186,000 in the form of 1.86×10^5? We used the rule that when we place a number into scientific notation form, the coefficient must be greater than one but less than 10. Since 186,000 is a positive number, we move the decimal point to the left and stop just before the last digit.

1 8 6 0 0 0 (Move 5 places and stop between the 1 and 8)

The number of times you move the decimal point becomes the power of ten or exponent part of the scientific notation form. In this case, the decimal place was moved five times to the left. Thus, the exponent becomes 10^5.

For small numbers, we move the decimal point to the right, to just after the first nonzero digit, and we use a negative exponent. As mentioned above, the mass of a typical iron atom is 0.0000000000000000000035 grams. To place this number into scientific notation, move the decimal point to the right and stop after the 3 in front of the 5. You will notice that to accomplish this you move the decimal point 21 times to the left:

0.0 0 0 0 0 0 0 0 0 0 0 0 0 0 0 0 0 0 0 0 3 5 g

When you are finished, the number looks like this:

$$3.5 \times 10^{-21} \text{ g}$$

As you can see, extremely large and small numbers written in this condensed format are much more manageable and easier to use in calculations.

There is another advantage to putting numbers into scientific notation: you will always know how many significant digits are present in the number. The number of digits in the coefficient part of the number in scientific notation form equals the number of significant digits present in the number. For example, 3.5×10^{-21} contains two significant digits and 1.85×10^5 contains three significant digits.

APPLYING THE CONCEPT: EXERCISE 4.1

1. Place the following numbers in exponential notation.

   0.0000000678   _____

   17200   _____

   1,244,000   _____

   0.000008091   _____

2.   Write out in full the following numbers.

5.67×10^{-6} _____

9.32×10^{4} _____

1.055×10^{-4} _____

3.13×10^{5} _____

MEASURING MASS

Mass is the amount of matter that a substance contains. Do not confuse the concept of mass and weight. The **weight** of an object depends upon the force of gravity and varies with geographic location. For example, you would weigh one sixth of your earth weight if you were on the surface of the moon because the gravitational pull of the moon is one sixth that of the gravitational pull of the earth. An object weighs more the closer it is to the center of the earth and less the further it is above the surface of the earth. Mass does not depend upon gravity and is therefore constant or independent of geographic location. Although mass and weight have different meanings and applications, they are frequently used interchangeably in chemistry.

We often speak of "weighing" a chemical sample in the laboratory when we are actually "massing out" the object or determining its mass. The base unit for mass in the metric system is the gram. Small amounts of mass would be measured in milligrams or centigrams, whereas large amounts of mass would be measured in kilograms.

Mass is an important variable that is measured frequently in the laboratory and used in many chemical calculations. For example, we must measure the mass of an object to determine its density. Mass is a variable that also appears in equations used in determining specific heat, calories absorbed or released upon heating and cooling, and number of moles present in a chemical sample.

In the laboratory, you will use the analytical balance to mass out samples of chemicals or other substances. Analytical balances are easy to use because they are digital. You simply turn on the analytical balance, place the sample on the weighing pan of the balance, and record the numbers displayed. The analytical balance in your laboratory may display the mass to the tenths, hundredths, or thousandths place, depending on the model. Your laboratory instructor will demonstrate the proper use of the particular model of analytical balance used in your laboratory. The photograph below illustrates two types of laboratory balances used for massing out samples in the chemistry laboratory.

Figure 4.1 Two types of laboratory balances
Courtesy of Brinkman Instruments Co.

APPLYING THE CONCEPT: EXERCISE 4.2

Why do you weigh slightly less on a mountain top than you do at sea level?

MEASURING LENGTH

The base unit for measuring length in the metric system is the meter. The centimeter or millimeter is an appropriate unit for measuring small distances. The meter, dekameter, and kilometer could be used to measure large distances. More than likely, you will use a metric ruler or a meter stick to make simple measurements of lengths in the laboratory. It is important to take the time to study the way the metric ruler or meter stick is broken down into decimeters, centimeters, and millimeters.

You will be expected to measure the lengths of various objects in the laboratory and to record your measurements to the proper number of significant digits. If you examine the meter stick carefully, you will note that you can record a measurement to only two decimal places. Why is this so? Suppose you measure the length of a block of wood by using a metric ruler (Figure 4.2). How should you record the measurement?

We know for sure that the block of wood is 2 cm long. We know for sure that the block of wood is 2 cm long plus 5 millimeters. But as you can see, the length of the block of wood is greater than 2.5 cm but less than 2.6 cm. The smallest division on the metric ruler is the millimeter. Consequently, we record the measurement as 2.55 by estimating the last digit. The number of digits we record with our measuring device is limited to the smallest division on the measuring instrument. If you remember the definition of significant figures, you will understand immediately why we can record to only two decimal places when using the metric ruler or meter stick. Significant digits contain all the digits in a measurement that are known for certain plus the first digit you have to estimate. This is why it is important to study every measuring device you use in the laboratory and know what the smallest division represents numerically.

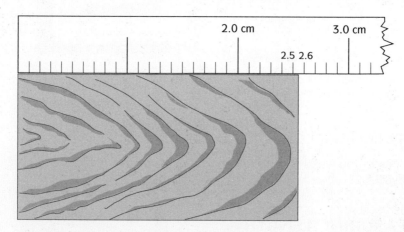

Figure 4.2 Measuring using a metric scale

APPLYING THE CONCEPT: EXERCISE 4.3

How are significant digits related to the measuring instrument used to take a measurement?

MEASURING VOLUME

In the laboratory chemistry students frequently measure the volume of liquid reagents needed for carrying out a chemical reaction. The basic unit of volume in the metric system is the liter. If you need a frame of reference, a liter is slightly larger than a quart. Most soda products are stocked in grocery stores in one liter and two liter containers. Small amounts of liquid volume are measured in milliliters. You should know from the prefix "milli" that 10^3 ml equals one liter.

One milliliter of liquid volume is also equal to exactly one cubic centimeter (cm^3). Doctors and nurses inject patients with medication using syringes calibrated in cubic centimeters. If a patient receives an injection of 5 cc's of lithium sulfate, he or she is receiving five cubic centimeters or five milliliters of the drug lithium sulfate.

In the laboratory, you will be using several types of instruments to measure the volume of liquids. Beakers, graduated cylinders, pipets, burets, and volumetric flasks are all instruments used in the measurement of volume. Beakers and graduated cylinders are used when we need only an approximate reading of the volume. Burets and volumetric flasks are used when we need to have a more exact reading of the volume. As with every measuring instrument you use in the laboratory, you should determine what the smallest division on the instrument represents to record the proper number of significant digits for the measurement. Figure 4.3 illustrates the various laboratory instruments used in measuring volume.

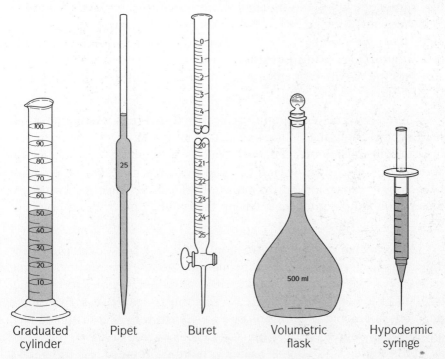

| Graduated cylinder | Pipet | Buret | Volumetric flask | Hypodermic syringe |

Figure 4.3 Some instruments used to measure volume

APPLYING THE CONCEPT: EXERCISE 4.4

Find a picture in your textbook or laboratory manual of a graduated cylinder and a buret. Which measuring device would allow you to record the greater number of significant digits when making a measurement of volume? How did you decide on your answer?

MEASURING TEMPERATURE AND HEAT

Temperature

A thermometer measures how hot or cold something is in degrees. That is, we use a thermometer to take a "temperature reading." The real question is how is temperature related to heat? Remember that adding heat to a substance causes the molecules of that substance to move faster. Molecules that have motion possess a special kind of energy known as **kinetic energy.** The amount of kinetic energy that a moving body possesses depends upon the mass of the body and its velocity. Mathematically, this relationship is expressed as $KE = MV^2/2$. When you a heat a substance, not all the molecules are moving with the same velocity. Some molecules are moving slowly whereas other molecules are moving very fast. Collectively, however, all the molecules have an average velocity. When you use a thermometer in the laboratory to take a **temperature** reading, what you are actually measuring is the average kinetic energy of the molecules of the substance being heated.

There are three temperature scales used in scientific study. You already know about the Fahrenheit scale, but it is not used in chemistry. Instead, the Celsius or centigrade scale is used along with the Kelvin scale. The Kelvin scale was developed primarily to study the effect of heat on the expansion and contraction of gases. Table 4.1 compares these three temperature scales using the boiling and freezing points of water as a reference.

How can we use the information in Table 4.1 to develop a set of equations to convert temperature measurements between these three scales? First, we notice that it takes 180 Fahrenheit degrees to be equal to 100 Celsius and Kelvin degrees. This must mean that a Fahrenheit degree is not as big as a Celsius or a Kelvin degree. As a matter of fact, a Fahrenheit degree is 5/9 as big as a Celsius degree (100/180 = 5/9), or a Celsius degree is 9/5 bigger than a Fahrenheit degree (180/100 = 9/5).

TABLE 4.1   THE THREE TEMPERATURE SCALES

	FAHRENHEIT	CELSIUS	KELVIN
Boiling Point	212	100	373
Freezing Point	32	0	273
Difference in degrees	180 (212 − 32 = 180)	100 (100 − 0 = 100)	100 (373 − 273 = 100)

Finally, we see that a correction factor of 32 degrees is needed to compensate for the difference in freezing point references in the Celsius and Fahrenheit scale. Putting all the information obtained from Table 4.1 into a set of mathematical expressions, we have the following:

1. $F = 1.8(C) + 32$

2. $C = \dfrac{(F - 32)}{1.8}$

3. $K = C + 273$

Now that you know where the numbers come from in the above equations, it should be easier for you to commit the formulas to memory and use them in temperature conversion problems.

APPLYING THE CONCEPT: EXERCISE 4.5

Most people are comfortable at a room temperature of about 74 degrees Fahrenheit. What is this Fahrenheit temperature in Celsius degrees and Kelvin degrees?

Heat

We often heat chemicals in the laboratory to cause a desired chemical reaction to occur. **Heat** is a form of energy that is related to the motion of atoms and molecules that make up a substance. The motion of the particles of a substance increases when we add heat. The heat eventually causes the substance to undergo a phase change. The substance may change from a solid to a liquid or from a liquid to a gas.

Suppose you have a cup of tea and a 20 by 40 in-ground swimming pool. First, you place a thermometer in the cup of tea and you get a reading of exactly 80°F. Next you place the thermometer in the swimming pool and the thermometer also reads exactly 80°F. Obviously, both the tea in the cup and the water in the swimming pool have the same temperature. But the swimming pool possesses much more heat than the cup of tea because it contains a greater quantity or mass of water. We now know that how much heat a substance possesses depends upon two factors: its temperature and its mass.

Traditionally, chemists have used the calorie as the unit of heat. The **calorie** is defined as the amount of heat required to raise the temperature of 1.0 gram of water through one degree Celsius. If we heat one gram of water from five degrees Celsius to fifteen degrees Celsius we have added 10 calories of heat to the water. Another unit for measuring all forms of energy is the joule. The **joule** is the official unit for expressing energy. Therefore, it is a good idea for you learn the relationship between calories and joules in case you need to convert back and forth between the two units. To convert from calories to joules and vice versa, use the following conversion factor:

$$1 \text{ calorie} = 4.184 \text{ joules}$$

There is a third factor that must be considered when determining how much heat an object possesses: specific heat. Pure substances vary in their ability to absorb and release heat per unit mass. We know from experience that metals heat up very quickly and also cool down very quickly. Water, on the other hand, absorbs heat very slowly but retains the heat and releases it very slowly. We say that water has a "high" specific heat and metals have a "low" specific heat. The amount of heat (in calories or joules) necessary to raise the temperature of one gram of that substance by one degree Celsius is called the **specific heat** of that substance. You need to memorize that the specific heat of water is 1.0 cal/g C or 4.184 J/g C. If you need to know the specific heat of other pure substances, look the values up in a chart. More than likely, you will find a table of specific heats in a chapter of your textbook or in an appendix in the back of your textbook.

APPLYING THE CONCEPT: EXERCISE 4.6

Find the chart listing the specific heat of various substances in your textbook. List the specific heat for the following substances:

aluminum _____

ice _____

We now know all the variables necessary to determine how many calories or joules of heat an object contains:

mass (m) in grams

temperature change $(T_2 - T_1)$ in Celsius degrees

specific heat (SH) in $\dfrac{cal}{g\ C}$ or in $\dfrac{J}{g\ C}$

Putting all these variables together into a simple algebraic equation, we come up with the **heat equation.** The amount of heat an object contains is equal to the specific heat times the mass times the temperature change:

$$Heat = SH \times m \times (T_2 - T_1)$$

If we use calories as the unit of heat, then all other units in the equation should cancel except calories. Let's do a quick unit check to verify this:

$$calories = \frac{calories}{\cancel{g}\ \cancel{Celsius}} \times \cancel{g} \times \cancel{Celsius}$$

As you can see, grams cancel with grams and degrees cancel with degrees, leaving the desired unit in calories.

Changes in energy always accompany chemical changes. Most frequently, the energy change is in the form of heat. When a chemical reaction occurs, heat is either released or absorbed. The amount of heat energy released or absorbed is directly related to the chemical energy stored within the chemical bonds of the substances reacting. Chemists cannot measure this chemical energy directly.

If the reaction is carried out in a calorimeter, the chemical energy can be measured by calculating its effect on temperature changes of water surrounding the calorimeter. This is the method used to determine the caloric content of foods. Have you ever wondered how food companies determine that a serving of cereal contains 130 calories or a chocolate Hershey bar contains 300 calories? Food chemists working in laboratories use a Bomb Calorimeter to explode weighed samples of foodstuff. The energy released as the food sample is vaporized is released into the surrounding water and causes a temperature change to occur. Using the heat equation, the chemists calculate the amount of calories contained in the food sample.

APPLYING THE CONCEPT: EXERCISE 4.7

How many calories are required to warm 1 kg of water from 4°C to 40°C?

USING MEASURED VARIABLES TO DETERMINE PHYSICAL PROPERTIES OF SUBSTANCES

In our discussion of measurement, we have seen that mass, length, volume, temperature, and heat are variables that are measured frequently in the chemistry laboratory. These measured variables can be used in various combinations to determine mathematically the physical properties of substances. Let's review some the most important physical properties of pure substances in which the above variables are used.

Density

Density contains two variables: mass and volume. Every pure substance has its own unique density. **Density** is the ratio of mass to volume. The equation for density is stated as follows:

$$\text{Density} = \frac{\text{mass}}{\text{volume}}$$

As you can see from the formula, the appropriate units for density would be g/cm^3 for a solid substance or g/ml for a liquid substance. Since gases usually have such large volumes, the appropriate unit for density is g/L. Density is often used by chemists along with other physical and chemical characteristics to identify unknown chemical substances.

Specific Gravity

Specific gravity is simply the ratio of the density of a pure substance compared to the density of water. Here is the equation for specific gravity:

$$\text{Specific Gravity} = \frac{\text{Density of pure substance}}{\text{Density of water}}$$

You need to know that the density of water is $1.0 \ g/cm^3$ or $1.0 \ g/ml$. Since the units for density are on both the top and bottom portions of the ratio, they cancel out during calculations. It should also be obvious that dividing any number by 1.0 gives you the exact same number. Therefore, the specific gravity of a pure substance is numerically equal to its density and has no units.

Specific Heat

Specific heat contains three variables: calories, mass, and temperature. As already defined, specific heat is the amount of heat needed to raise the temperature of one gram of a pure substance gains by one degree Celsius. The equation for specific heat is stated as follows:

$$\text{Specific Heat} = \frac{\text{cal}}{\text{g deg}}$$

Remember, specific heat is an important part of the equation in determining the heat (energy) content of chemical substances.

Heat Energy

From our previous discussion, you already know that the heat energy equation contains the variables mass, specific heat, and temperature, or:

Calories (heat energy) = specific heat × mass × temperature change

PROBLEM-SOLVING STRATEGIES IN MEASUREMENT

In chemistry, measured variables are used in carrying out calculations to find answers to problems. You will remember that in Chapter 2 under the section Taking Control of Your Chemistry Homework we discussed problem-solving strategies. Please review this section and become familiar with the problem-solving model presented.

As stated in Chapter 2, almost all beginning chemistry problems can be solved by using a conversion factor, ratio and proportion, or a simple algebraic equation. Let's briefly review these three approaches.

Conversion Factors

Conversion factors are used frequently in measurement. They are used to convert one unit of measurement to an equivalent unit of measurement. For example, a conversion factor could be used to convert kilograms to pounds, inches to millimeters, and meters to centimeters. A conversion factor is simply a ratio of two forms of a unit that are equivalent. It is important to recognize that conversion factors can always be written in two forms. But only one of the forms will allow you to get the right unit and answer in the solution pathway you set up to solve the problem. Some examples of conversion factors are illustrated in Table 4.2.

TABLE 4.2   CONVERSION FACTORS

FORM 1	FORM 2
$\dfrac{365 \text{ days}}{1 \text{ year}}$	$\dfrac{1 \text{ year}}{365 \text{ days}}$
$\dfrac{12 \text{ inches}}{1 \text{ foot}}$	$\dfrac{1 \text{ foot}}{12 \text{ inches}}$
$\dfrac{2.2 \text{ lbs}}{1 \text{ kg}}$	$\dfrac{1 \text{ kg}}{2.2 \text{ lbs}}$
$\dfrac{100 \text{ cm}}{1 \text{ m}}$	$\dfrac{1 \text{ m}}{100 \text{ cm}}$

APPLYING THE CONCEPT: EXERCISE 4.8

Express the following statements as conversion factors. Write two forms for each conversion factor.

1. 4 pints equal 1 quart
2. 3600 seconds equals 1 hour
3. 1 kilogram equals 1×10^6 milligrams

In problem solving, how do you know which form of the conversion factor to use? The answer is quite simple. Using the problem-solving model given in Chapter 2, you identify the unknown variable and the unit it requires. Then, you pick the conversion factor that will allow you to cancel out unwanted units and end up with the correct unit. Let's revisit the example problem given in Chapter 2 on pages 20–21 to illustrate this point.

A chemistry student needs 140 millimeters of aluminum wire for an experiment. The student has only a ruler calibrated in inches. How many inches of aluminum wire does the student need to cut to provide the needed 140 millimeters of wire?

1. Write down the problem solving model just as it is shown below.
   Known or Given → Connections → Goal or Unknown

2. Write down all given variables by name and numerical values under the Known or Given column and identify the unknown variable or goal state under the Goal or Unknown column.

   Given or Known → Connections → Goal or Unknown

   140 mm Al wire   inches of Al wire

3. Identify the connection(s) needed to define the solution pathway to obtain the desired answer.

   Given or Known → Connections → Goal or Unknown

   140 mm Al wire $\dfrac{2.54\ cm}{1\ in}$ or $\dfrac{1\ in}{2.54\ cm}$ inches of Al wire

   $\dfrac{10\ mm}{1\ cm}$ or $\dfrac{1\ cm}{10\ mm}$

4. Set up the problem in a series of steps using the identified connection(s) and check to see if the problem will yield the desired unit.

   $140\ mm\ Al \times \dfrac{1\ cm}{10\ mm} \times \dfrac{1\ in}{2.54\ cm} = \quad in$

5. Carry out the mathematical operations identified in the solution pathway and express the numerical part of the answer to the correct number of significant digits. Make sure your answer contains the right unit as well as the correct number of numerical digits.

   $140\ mm\ Al \times \dfrac{1\ cm}{10\ mm}$ or $\dfrac{1\ in}{2.54\ cm} = 5.5\ in\ Al$

Pay particular attention to step 3. Notice that two conversion factors were identified and two forms of the conversion factors were written to use in the solution pathway for this problem. Now, notice in step 4 that of the two possible sets of conversion factors, only one set will allow you to end up with the desired unit (inches) after you complete the solution pathway. If you had chosen the other set of conversion factors for this problem, you would have ended up with:

$$140 \text{ mm} \times \frac{10 \text{ mm}}{1 \text{ cm}} \times \frac{2.54 \text{ cm}}{1 \text{ inch}} = 3556 \text{ mm}^2/\text{inch}$$

As you can see, the units make no sense and certainly are not the correct units the problem calls for. You will save yourself a lot of time and frustration if you will remember to check units first and do calculations second in solving chemistry problems.

APPLYING THE CONCEPT: EXERCISE 4.9

Solve the following problem using the problem-solving model above. Write down each step and show the appropriate steps or calculations needed.

The speed limit on most freeways in the United States is 65 miles/hour. What speed does this represent in meters/sec? (1 mile = 1.6094 km)

Ratio and Proportions

Many chemistry problems can be solved by using a simple ratio and proportion approach. The key to solving these types of problems in chemistry is to determine if two variables are proportional to each other. Ratio and proportion problems have the following form:

$$\frac{A}{B} = \frac{C}{D}$$

It is important to remember in working ratio and proportion problems that the units in the numerator and denominator must be consistent. In clinical chemistry, ratio and proportion are useful in calculating dosage factors. Consider the following problem:

If 2 mg of a drug is administered for each lb of body weight, how many mg of the drug should be administered to a 7.5 lb infant?

Solution: $\dfrac{2 \text{ mg drug}}{\text{lb}} = \dfrac{x}{7.5 \text{ lb}}$

$$x = \frac{2 \text{ mg drug} \times 7.5 \text{ lbs}}{\text{lbs}}$$

$$x = 15 \text{ mg drug dosage}$$

In this problem, drug dosage and body weight are proportional to each other. As the body weight increases, the dosage weight increases. Hence, ratio and proportion is the most appropriate problem-solving approach to use in this situation.

Simple Algebraic Equations

Many chemistry problems are solved by using defined algebraic equations. Some examples include finding the pH of solutions, the gas law equations, calculating enthalpy, and standard heats of formation. Every equation you are expected to use in the solution of chemistry problems will be explained in detail in your chemistry textbook. Do not expect, however, the equation to be in the correct form you will need for problem solving. It is important to read the problem carefully, to identify the equation that relates to the problem, and to rearrange the equation for the unknown term if needed.

What about applying thinking skills to problem solving? Solving chemistry problems requires critical thinking and the use of higher-order thinking skills. To solve a chemistry problem, you must comprehend the problem (comprehension level). Comprehending the problem means that you understand the problem and recognize the goal. Recognizing the goal of the problem might require that you outline or diagram the problem (analysis level). Finally, you solve the problem by setting up the solution pathway using relevant information given in the problem (application level). As you can see, solving chemistry problems requires investing the mental energy to plan out a solution by critically thinking about information given in the problem. After comprehending the problem and identifying relevant information, you apply a solution pathway that leads to the goal or desired state of the problem.

USING YOUR CALCULATOR

A calculator is a necessary investment for beginning chemistry students. It is important that you purchase a "scientific calculator." A scientific calculator doesn't have to be expensive; many good models cost less than $15.00. What is important is that the calculator that you purchase has the types of keys listed in Table 4.3.

Placing numbers into exponential notation is an important calculator skill in working measurement problems. Take a moment to examine your calculator and find the exponent key. Practice placing the following numbers into the calculator using the exponent key by following the steps in Table 4.4.

Notice that you do not key in the number 10 when you are using the calculator to place numbers into scientific or exponential notation. The "exp key" places the calculator in the exponential mode automatically. Many beginning chemistry students key in exponential numbers incorrectly by including "10" in the keystroke sequence. This sequence results in an answer that is off by a power of 10.

Suppose you keyed in the number 6.02×10^{23} in the following manner:

$$6.02 \times 10 \text{ [exp key] } 23$$

TABLE 4.3   IMPORTANT KEYS ON A CALCULATOR

KEY	FUNCTION
Exponent key (exp or ee)	Places numbers into exponential notation
y^x key	Raises a number to a certain power
Logarithm (log) key (natural and base 10)	Gives the natural log or base 10 log of any number
Inverse key (2nd)	Used in conjunction with other keys to allow one key to have a dual function. For example putting a number into the calculator and hitting the log key displays the log of the number. Putting a number representing a log into the calculator and pressing (2nd) (log) displays the number the log represents.
[+/−] key	Used to change a positive number to a negative number and vice versa.

The calculator window would display 6.02 24 instead of the correct form 6.02 23. Remember, do not key in the number 10 as part of the keystrokes when you are using your calculator to place numbers into scientific or exponential notation.

TABLE 4.4   USING THE CALCULATOR

NUMBER	CALCULATOR KEY STROKES	CALCULATOR DISPLAYS
6.02×10^{23}	6.02 [exp key] 23	6.02 23
4.85×10^{-8}	4.85 [exp key] 8 [+/−key]	4.85 − 08
-1.92×10^{11}	1.92 [+/−key][exp key] 11	−1.92 11

APPLYING THE CONCEPT: EXERCISE 4.10

Complete the following using your calculator:

$2.14 \times 10^{-14} \times 1.77 \times 10^{4} =$

TERMS TO KNOW

calorie

density

exponential notation

heat

heat equation

joule

kinetic energy

mass

scientific notation

specific gravity

specific heat

temperature

weight

CHAPTER 5

THE HEART OF CHEMISTRY: Organization of Matter

GETTING FOCUSED

- *How do you think matter that comprises the earth is organized?*

- *Do you learn and remember concepts better when you can visually see how concepts are related to each other?*

- *Do you know what a concept map is? Have you ever used a concept map in learning complicated concepts?*

- *How are atoms of all the elements similar? How are they different?*

- *Why is the Periodic Table so important to chemists? What are some of the different types of chemical information you can extract from the Periodic Table?*

- *Why do the atoms unite to form compounds?*

TAKING INVENTORY

Learning Inventory 2 is designed to evaluate your knowledge and perceptions about the composition of matter and atomic structure. Don't worry about the "rightness" or "wrongness" of your answers at this point. The most important thing is to think about each question and write down what you believe or know about the topic addressed in each question.

When we learn a new concept, we anchor new information about that concept onto existing information. It is important to clear up any misconceptions that you might have about the concept of the organization of matter as you study this chapter and the related topic in your textbook. As you answer these questions, don't use your textbook or another source as a reference. The purpose of this exercise is for you to evaluate your existing knowledge about the topic. After reading and studying this chapter along with your textbook, you will be given another opportunity to evaluate your knowledge about the composition of matter. It would be useful to compare your responses before and after studying this chapter to determine if your knowledge and viewpoints about the composition of matter have changed.

Learning Inventory 2

Read each question carefully and provide a short response.

1. Matter can be grouped into three categories: elements, compounds, and mixtures. How do these three categories of matter differ from one other?

_____

_____

_____

_____

_____

_____

2. All matter is made up of tiny, invisible particles called atoms. What is your interpretation of the meaning of the word "atom"?

_____

_____

_____

_____

_____

3. What do you know about protons, neutrons, and electrons?

_____

_____

_____

_____

4. Walk into any chemistry lecture room or laboratory room and you will see a chart of the Periodic Table hanging on the wall. In your opinion, why is the Periodic Table so important to the chemist and the chemistry student?

_____

_____

_____

_____

5. How are atoms and ions similar? How are they different?

_____

_____

_____

_____

_____

6. There are literally millions of chemical compounds that occur naturally and are synthesized in chemical laboratories. Briefly summarize your thoughts on how atoms and ions participate chemically in forming compounds.

_____

_____

_____

_____

_____

MAPPING AND ORGANIZING THE STRUCTURE OF MATTER

Graphic organizers and concept maps are useful tools to visualize the relationship among concepts. Chemistry concepts do not exist in isolation. Therefore, quite frequently you will study several chemical concepts that are "linked" together. Networks are graphic organizers or concepts maps in which the relationship among several concepts are visualized. Research has shown that the use of network organizers helps students improve their understanding of scientific concepts.

What are the type of "links" or relationships that are used to connect concepts together in a network organizer? Diane Halpern, in her book *Thought and Knowledge: An Introduction to Critical Thinking* citing the work of Holley and Dansereau, identifies links used in developing network organizers. These links are shown in Table 5.1.

Let's use the links described in Table 5.1 to construct a network organizer for the organization of matter. A good starting point would be to generate a list of key terms and concepts about the organization of matter and then link them together into a visual map. What do we already know about the composition of matter?

In chemistry, matter is organized into three broad categories: elements, compounds, and mixtures. Each of these three categories has its own unique set of physical and chemical properties. Elements and compounds are homogeneous substances. Homogeneous substances have a uniform or constant composition. There are 109 known elements known today. Of these 109 elements only 88 occur naturally, the rest are manmade elements. Chemical compounds are the results of the

TABLE 5.1   LINKS FOR DEVELOPING NETWORK ORGANIZERS

TYPE	EXAMPLE	KEY WORDS
Part of link X is a part of Y	atom -**p**- electron	is a part of is a segment of is a portion of
Type of/example of link X is a type of Y	bond -**t**- ionic	is a type of is in the category is an example of is a kind of
Leads to link X leads to Y	chemical changes -**l**- energy changes	leads to results in causes produces
Analogy link X is like Y	electrons around an atom -**a**- bees around a beehive	is similar to is analogous to is like corresponds to
Characteristic link X is a characteristic of Y	metal -**c**- conducts electricity	is characterized by feature is property is trait is aspect is attribute is
Evidence link X is evidence that Y occurred	chemical reaction - **e** - color change	indicates is illustrated by is demonstrated by supports documents is proof of confirms

chemical combination of the atoms of these elements. Mixtures are heterogeneous substances with a varying composition of elements and compounds and can be separated by ordinary physical means such as filtering and distillation.

All matter is composed of atoms. Atoms in turn are composed of subatomic particles. The three most important subatomic particles are protons, neutrons, and electrons. Elements are composed of a single type of atom whereas compounds are composed of two or more elements whose atoms are chemically united in a definite proportion. When two or more atoms chemically unite in a definite ratio, they form molecules such as H_2O and $C_6H_{12}O_6$. Also, the atoms united in chemical compounds can either share electrons by way of a covalent bond, or transfer electrons forming positive and negative ions, which results in an ionic bond.

What about solutions? Chemists and chemistry students in the laboratory are constantly making solutions and mixing solutions together to bring about chemical changes. Solutions are unique in that they are homogeneous mixtures. Solutions are prepared by placing a solute (the substance being dissolved) into a solvent (the substance doing the dissolving). You can recover both the solute and the solvent in a solution by evaporating the solvent.

There are many other important concepts related to the organization of matter. The atoms that comprise a given element are not always identical. For example, analysis of a pure element might reveal that it is composed of three different forms of atoms that are identical in every respect expect mass. Such types of atoms are called isotopes. **Isotopes** are atoms of an element that differ in the number of neutrons contained in their nucleus, hence they vary in mass. The isotopes of several elements such as iodine and radium are extremely important in medicine because they are radioactive. They are used in the diagnosis and treatment of cancer and many other diseases.

As you can see, there are a lot of connections or linkages that unite the components of matter into a unified concept which we will call "the organization of matter." What we need to do to better understand how these components "fit" together is to draw a graphic organizer for the concept of the organization of matter. There is a lot of truth in the old saying: "a picture is worth a thousand words." You have probably never attempted to draw a graphic organizer in studying, so you can refer to Figure 5.1 as an example of one for studying this concept.

A graphic organizer is a good method for you to see the "big picture" by analyzing how all the small parts fit together to create the total concept. You may find, however, that you do not know enough factual information about one of the components or parts of the concept. Suppose when you finish studying the concept map for the organization of matter you realize that you don't fully understand the relationship of

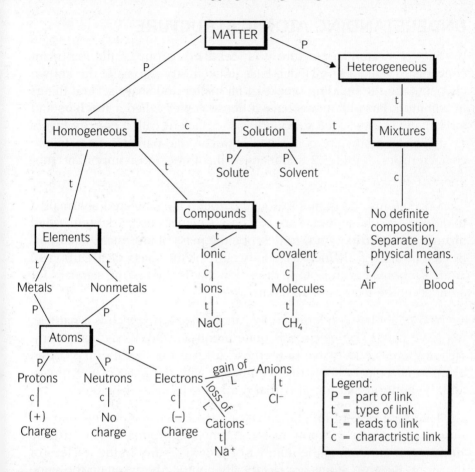

Figure 5.1 A graphic organizer for studying the organization of matter

isotopes to atoms. This is where your textbook and lecture notes come to the rescue. Read and review the section in your textbook and your lecture notes concerning isotopes. Stop and summarize the relationship in writing, keeping in mind the type of link that connects the concept of atoms and isotopes used in the concept map.

We will be using graphic organizers a lot in the upcoming chapters of this study skill textbook. As you study topics in your textbook in which several concepts are being discussed, it is a good idea for you to draw a graphic organizer or concept map to help visualize the relationships or connections among the concepts under study.

UNDERSTANDING ATOMIC STRUCTURE

In Chapter 1, atomic structure was identified as one of the important concepts to build a knowledge base in chemistry. Atoms of the known elements are the building blocks of all matter and share several things in common. First, all atoms have a dense center called a **nucleus** and a relatively empty space surrounding the nucleus called **energy levels** or shells. All atoms are composed of subatomic particles with unique characteristics. Table 5.2 summarizes the three most important subatomic particles.

Since atoms and their subatomic particles are so incredibly small, a traditional weight system is not practical. Chemists used a system called **atomic mass units** (amu) to describe the mass of atoms and their particles. An amu is defined as one-twelfth of the mass of a carbon-12 atom. Carbon-12 is an isotope of carbon that is used as the standard for comparing the masses of all atoms.

You might be wondering, how can an electron, which is a particle, not have mass? The electron is quite unique in that it has both "particle" properties and "wave properties." It turns out that the mass of an electron is so incredibly small compared with the mass of the proton and the neutron that for all partial purposes, its mass is ignored.

There are several numbers that are associated with atoms that are important for you to know and use when appropriate. The **atomic number** of an atom is the number of protons found in the nucleus of the atom. Because atoms are electrically neutral, they contain the same number of positive and negative charges. Consequently, the atomic number also tells you how many electrons are found around the nu-

TABLE 5.2   THE THREE MOST IMPORTANT SUBATOMIC PARTICLES

PARTICLE	SYMBOL	LOCATION	CHARGE	MASS
Proton	p or p$^+$	Nucleus	1+	1 amu
Neutron	n or n^0	Nucleus	none	1 amu
Electron	e$^-$	outside nucleus in energy levels	1−	0

cleus. The **mass number** of an atom is the total number of protons
and neutrons found in the nucleus of an atom. An atom of gold has 79
protons and 118 neutrons in its nucleus. Therefore, gold atoms have a
mass number of 197. The **atomic weight** (mass) of an atom is the
weighted average of all the naturally occurring isotopes of a given element. This explains why most of the atomic weights (mass) listed in
charts in chemistry textbooks contain decimal places.

APPLYING THE CONCEPT: EXERCISE 5.1

Use the Periodic Table in your textbook to answer the following:
1. How many protons, neutrons, and electrons are found in an aluminum atom?
2. What is the mass number for chlorine?
3. Which element has an atomic number of 11?
4. Which element has an atomic weight of 40.078 amu?

ELECTRON ARRANGEMENT

It is the arrangement of electrons around the nucleus of an atom that
determines the atom's "reactivity" or tendency to combine chemically
with other atoms to form compounds. Electrons are located outside the
nucleus of the atom in regions of space called energy levels or shells.
Electrons always occupy the lowest energy shell available to them. The
first energy shell is closest to the nucleus and has the lowest energy
value. It occupies a relatively small region around the nucleus and contains only two electrons. Atoms having more than two electrons must
use energy shells further out from the nucleus. Atoms with large atomic
numbers, and hence large numbers of electrons, obviously have many
energy shells. Beginning chemistry students usually work with atoms
whose electrons occupy the first four energy shells. The way electrons
in an atom are arranged in energy shells and sub shells around the nucleus is called the **electron configuration** of the atom. Karen Timberlake in *Chemistry: An Introduction to General, Organic, and Biological
Chemistry,* Fifth Edition, uses a very creative analogy called the "Electron Hotel" to show the arrangement of electrons in energy shells and

sub shells around an atom. Figure 5.2 represents Dr. Timberlake's "Electron Hotel." Dr. Timberlake describes her Electron Hotel analogy as follows:

> *The electron arrangement of an atom gives the number of electrons in each shell. We might imagine electron shells as floors in a hotel. The ground floor fills first, then the second floor, and so on. In Figure 5.2, two electrons are placed in shell 1, the lowest energy shell. The next eight electrons go into shell 2. Both shell 1 and 2 are now filled. Shell 3 initially takes eight electrons, it then stops filling for a while, even though it is capable of holding more electrons. This break in filling is due to an overlapping in the energy value of shell 3 and shell 4. At this point, shell 4 takes the next two electrons.*

The electron arrangement described here is a very simplistic model. Electrons are also located in orbitals of a given energy value and molecular shape within the principle energy shells. These orbitals have traditionally been represented by the letters *s, p, d,* and *f.* If you are re-

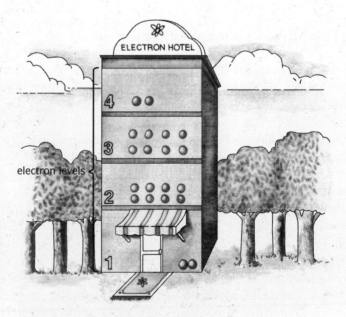

Figure 5.2 The Timberlake Diagram. From *Karen Timberlake, Chemistry: An Introduction to General, Organic, and Biological Chemistry*, 5th ed.

quired to use orbital notation in "writing out" the electron configuration of the electrons in a given atom you will need to know the kind (s, p, d, f), and number of orbitals contained within the principle energy shells (levels). Although orbital notation might seem a bit difficult in the beginning, you will soon find that a pattern (algorithm) exists that you can apply to any atom. If writing out electron configurations for atoms is an important skill for you to know, study the section in your textbook carefully. Practice writing out the electron configuration of the first 20 elements (atomic numbers 1 through 20).

It is most important that you be able to identify the outermost energy shell of an atom and the number of electrons located in that shell. Knowing the number of **valence** (outermost) electrons in a particular atom and the energy shell in which they are located will allow you to predict the types of compounds an atom will form with other atoms and the formula for the resulting compound. Later on in this chapter you will see how a working knowledge of the Periodic Table will allow you to determine this information.

APPLYING THE CONCEPT: EXERCISE 5.2

1. Which energy shell or level is the outermost energy level for an atom of magnesium?
2. How many electrons are in the outermost energy level or shell of an atom of magnesium?
3. Using Timberlake's hotel analogy model, how would you show the electron arrangement for an atom of chlorine?

UNDERSTANDING THE PERIODIC TABLE

By now, you probably have had the opportunity to become familiar with the Periodic Table located on the inside cover of your chemistry textbook. The Periodic Table is an organizational chart of the known elements based on the concept of **chemical periodicity.** Through much trial and error, it was discovered that when the elements are arranged in rows by increasing atomic numbers, physical and chemical properties of the elements reoccur in repeating patterns. The rows of elements in the Periodic Table are known as **periods** or **series,** the columns containing elements are referred to as **groups** or **families.**

Open your textbook to the inside cover to the Periodic Table, or find a copy of it in the appropriate chapter of your textbook. Take a few minutes to get a "feel" of the layout of the table. There are several variations used by textbooks in the numbering system for the columns of the Periodic Table. A newer version numbers the columns of the table from 1 to 18. Most textbooks however, still use the A and B system of numbering the columns. The "A" columns, IA to VIIIA, are the regular elements whose outermost electron arrangement are consistent. The "A" columns make up the **representative elements.** This means that they always lose or gain a definite number of electrons during compound formation. The "B" column of elements are called the **transition elements.** These elements are not consistent in their exchange of outermost electrons during compound formation. For example, sometimes when copper (Cu) atoms form compounds with nonmetallic atoms, the copper atoms lose one electron and form 1$^+$ ions. But, copper atoms are also capable of losing two electrons during compound formation resulting in the formation of copper 2$^+$ ions. Since many of the transition metals form large numbers of important compounds, it is a good idea to memorize a few of the most common transition metals and the charges on the ions they form.

The most remarkable feature of the Periodic Table is that elements located in the same column or family have similar chemical properties. For example, all the elements in Group I (the alkali metals) combine with chlorine (Group VII) in a 1:1 ratio to form a class of compounds called **salts.** You are familiar with common table salt, sodium chloride. The chemical formula for sodium chloride is NaCl. If your doctor told you to try and restrict your sodium intake, you might switch to "light salt," which is potassium chloride. The chemical formula for potassium chloride is KCl. Note that in both NaCl and KCl the combining ratio of the two metals to chlorine is 1:1.

Why do elements in the same column of the Periodic Table exhibit similar chemical characteristics? The answer, of course, is electron distribution. Obviously, atoms in the same column of the Periodic Table do not have the same number of electrons. What these elements do have in common, however, is the same number of electrons in their outermost (valence) energy shell. All the elements in Group IA have one valance electron in their outermost shell. Elements in Group VIIA have seven electrons in their outermost energy shell. Atoms of elements with identical number of valence electrons behave similarly in their chemical reactivity.

It is very easy to determine the number of valence (outer shell) electrons for the "A" elements in the Periodic Table. The number of valence electrons is exactly equal to the Group number of the column in which the element is located. Suppose you wanted to know how many valence electrons are in an atom of aluminum (Al). Looking at the Periodic Table we see that aluminum is located in Group IIIA (3A). Therefore, aluminum atoms have three valence electrons. Similarly, calcium, located in Group IIA, has two valence electrons and chlorine, located in Group VIIA, has seven valence electrons. As already stated, knowing the number of valence electrons for a particular atom is important in determining how it will bond with other atoms and in predicting the formula for the resulting compound.

It is not the intent of this study skill textbook to teach you all the content information you need to know about the Periodic Table. You will need to study the section on the Periodic Table carefully in your textbook and in your lecture notes. Remember, the Periodic Table is a "tool" for you to use to extract important information about the elements, and to make predictions about their behavior. After reading and studying carefully the sections in your textbook related to the Periodic Table, can you use the Periodic Table to perform the following tasks?

- Locate the "A" or representative elements and the transition "B" elements.

- Separate the elements into three categories: metals, nonmetals, and metalloids.

- Find the alkali earth metals.

- Find the alkaline earth metals.

- Find the halogens.

- Locate the noble gases.

- Locate the atomic number for the atoms of the elements.

- Locate the atomic mass (weight) of the atoms of the elements.

- Determine the mass number of the atoms of the elements.

- Determine the number of proton, neutrons, and electrons in the atoms of the elements.

- Determine how many energy shells an atom of an element contains.

- Determine the number of valence electrons for the atoms of the "A" or representative elements.

- Determine the charge on the ion formed by the atoms of the "A" or representative elements.

- Write the formula for compounds formed by the combination of metallic atoms and nonmetallic atoms for the "A" group or representative elements.

- Predict how atomic radius (size) of the atoms changes within a column and across a period of the Periodic Table.

- Predict how the ionization energy of the atoms of the elements changes within a column and across the periods of the Periodic Table.

As you can see from the above list, the Periodic Table contains a wealth of information about the physical and chemical properties of the elements. If you take the time to learn how to fully use the table as an "information tool" it will pay off many times in your understanding of chemistry.

UNDERSTANDING THE FORMATION OF CHEMICAL COMPOUNDS

Most of the elements that are represented in the rows and columns of the Periodic Table rarely occur alone in nature. Almost always, elements are found in chemical combination with other elements in the form of compounds. Review the concept map for the organization of matter to see the relationship between atoms, elements, and compounds.

What is the chemical driving force for atoms and ions to unite to form chemical compounds? One of the most fundamental laws of nature is that all physical and chemical processes tend to proceed in a direction that leads to lower energy and higher stability. Objects above the surface of the earth are more energetic and less stable. When these objects fall and strike the earth, they are less energetic and more stable. What electron arrangement gives an atom or ion the greatest stability or lowest energy state? Uncombined atoms that do not have a filled valence shell or an octet of valence electrons are chemically unstable. These uncombined atoms seek other atoms to exchange or share electrons in order to obtain a filled valence shell or an octet of valence electrons. When atoms have a filled valence shell or eight valence electrons they enter a lower energy state and consequently, become more stable. From your study of the Periodic Table you already know that the "noble gases" are the most stable of the elements and under normal chemical conditions do not combine with atoms of other elements. The first noble gas, helium, has a filled first energy level that imparts high stability. All of the other noble gases have eight electrons in their valence shell. The way atoms gain, lose, or share electrons is attributed to their attempts at achieving the same number of valence electrons as the noble gas closest to them in the Periodic Table. This is often referred to as the **octet rule.** Those elements that do not have eight valence electrons seek other atoms to lose, gain, or share electrons with in order to end up with an octet of electrons.

When atoms react with other atoms or ions, only the outermost (valence) electrons are involved. It is the movement of valence electrons from one atom to another, or the sharing of valence electrons between two atoms that causes a compound to be formed. As a general rule, metallic atoms lose electrons to achieve an octet and form positive ions (cations). The following equations illustrate the loss of electrons by metallic atoms of the representative (A) elements:

$$Li \rightarrow Li^{1+} + e^-$$

$$Na \rightarrow Na^{1+} + e^-$$

$$Mg \rightarrow Mg^{2+} + 2e^-$$

$$Ca \rightarrow Ca^{2+} + 2e^-$$

$$Al \rightarrow Al^{3+} + 3e^-$$

Make sure you fully understand the following points about the above equations representing ion formation by metallic ions:

- The charge on the resulting metallic ion is the same as its group number. The charge on Li and Na is plus one. The group number for Li and Na is IA. The charge on Al is plus 3. The group number for Al is IIIA.

- All of the above atoms do not have a full valence shell or a stable octet. Therefore, all the above atoms are not chemically stable. The electron arrangement for the atoms are:

Li: 2,1
Na: 2,8,1

Mg: 2,8,2
Ca: 2,8,8,2
Al: 2,8,3

- Only the electrons in the outermost (valence) shell are given up by the metallic atoms

- The resulting ions formed by the atoms are stable because each one has either a filled valence shell or a stable octet. The electron arrangement for the ions are:

Li^{1+}:   2
Na^{1+}:   2,8
Mg^{2+}:   2,8
Ca^{2+}:   2,8,8
Al^{3+}:   2,8

What happens to the electrons lost by the metallic atoms in forming positive ions? Free electrons just simply do not exist in nature. Therefore, there must be another atom willing to accept the electrons lost by a metallic atom. Nonmetallic atoms gain electrons lost by metallic atoms, and form negative (anions) during compound formation. The following equations illustrate nonmetallic atoms gaining electrons and forming negative ions:

$$N + 3e^- \rightarrow N^{3-}$$
$$O + 2e^- \rightarrow O^{2-}$$
$$S + 2e^- \rightarrow S^{2-}$$
$$F + 1e^- \rightarrow F^{1-}$$
$$Cl + 1e^- \rightarrow Cl^{1-}$$

Again, the most important points to consider when interpreting the full meaning of the above equations are:

- The charge on the resulting nonmetallic ion is not the same as its group number. However, if you take the group number of the element and subtract the number from eight, it will give you the number of electrons needed by the nonmetallic atom to achieve a stable octet. For example, nitrogen is in Group VA. This means nitrogen has five valence electrons. Nitrogen needs three more (8 - 5 = 3) electrons to have a stable octet.

- None of the above nonmetallic atoms are stable because all lack a stable octet of electrons in their valence shells. The electron arrangement for the above atoms is:

N: 2,5
O: 2,6
S: 2,8,6
F: 2,7
Cl: 2,8,7

- The electrons gained by the nonmetallic atoms go only into the outermost energy shells of the atoms.

- The resulting negative ions formed when the nonmetallic atoms gain electrons are stable because each obtains an octet of valence electrons. The electron arrangement of the resulting ions is:

N^{3-}:   2,8
O^{2-}:   2,8
S^{2-}:   2,8,8
F^{1-}:   2,8
Cl^{1-}:   2,8,8

I hope you now have a better understanding of why it is important for you to be able to determine the number of valence electrons an atom possesses and the number of electrons needed in order to obtain an octet. Writing correct formulas for compounds is "linked" to the concept of the octet. Writing formulas for compounds is a prerequisite skill for balancing equations and working chemical problems dealing with the balanced equation.

CHEMICAL COMPOUNDS AND THEIR FORMULAS

There are an incredibly vast number of chemical compounds that occur naturally, and an equally large number that are synthesized in chemical laboratories throughout the world. Each and every one of these chemical compounds has associated with it a chemical formula that represents the kind and number of atoms and ions contained in the compounds. For example, the common substance baking soda is a chemical compounds composed of sodium ions and hydrogen carbonate (bicarbonate) ions. The chemical formula for baking soda is $NaHCO_3$. You already know about sodium ions. Sodium ions (Na^{1+}) are formed when sodium atoms, located in Group IA of the Periodic Table lose one valence electron. The hydrogen carbonate ion (HCO_3^{1-}) is a special type of ion called a **polyatomic ion.** You won't find polyatomic ions in the Periodic Table. They consist of several atoms bonded together and carry a charge. Find the chart listing the polyatomic ions, their formulas, and charges in your textbook. It is important that you be able to recognize a polyatomic ion by name and by its formula for these are used frequently in writing formulas for compounds and in giving names to chemical compounds.

Ionic Compounds

Ionic compounds are composed of crystals. These crystals in turn are composed of positive and negative ions arranged in a regular fashion. The electrostatic attraction between the positive and negative ions creates the ionic bond that holds the crystalline structure together. It is easy to recognize an ionic compound. Ionic compounds will always be composed of:

- A metallic ion and a nonmetallic ion such as KBr, CaO, and ZnS

- A metallic ion and a negative polyatomic ion such as $Al_2(SO_4)_3$ or Li_2CO_3

- A positive polyatomic ion and a negative polyatomic ion such as NH_4NO_3

- A positive polyatomic ion and a negative monatomic ion such as NH_4Cl

It is important that you are able to name ionic compounds correctly if you are given their formulas. The rules for naming ionic compounds are very straightforward, so the section in your textbook concerning naming ionic compounds should be sufficient. Spend time reviewing the chart in your textbook containing the names, formulas, and charges for the polyatomic ions. Many instructors insist that students memorize the polyatomic ion list. Other instructors allow students to refer to the list when naming and writing formulas for compounds. You will find, however, that after repeated use you will automatically commit to memory the names and formulas for many of the most commonly used polyatomic ions.

In writing formulas for ionic compounds the most important points to keep in mind are:

- The positive ion is always written first in the formula for an ionic compound.

- The overall charge on an ionic compound must be zero. All chemical compounds are electrically neutral. The sum of the charges on the positive (metallic) ions must be equal to the sum of the charges on the negative (nonmetallic) ions.

- Subscripts are used to indicate the number of positive and negative ions in combination needed for electrical neutrality. For example, in the compound Al_2O_3 two aluminum ions each with a charge of plus three are needed to combine with three oxide ions each with a charge of minus two for the overall charge on the compound to sum to zero.

- Parentheses are used around a polyatomic ion when it is taken more than once in the formula of an ionic compound. Examine the formula for aluminum nitrate $Al(NO_3)_3$. In this compound three nitrate ions each with a charge of minus one are in combination with one aluminum ion with a charge of plus one.

Make sure you spend sufficient time studying the rules and examples in your textbook on writing formulas for compounds. As a final review, and to check your formula writing skills, complete the exercises that follow.

APPLYING THE CONCEPT: EXERCISE 5.3

Give the formula and charge for the ion formed by the following atoms:

Lithium	Li^{1+}
Potassium	
Magnesium	
Barium	
Aluminum	
Oxygen	
Sulfur	
Fluorine	
Iodine	

APPLYING THE CONCEPT: EXERCISE 5.4

Write the formula and charge for the following polyatomic ions:

Ammonium	
Nitrate	
Nitrite	
Carbonate	
Hydrogen carbonate	
Phosphate	
Chlorate	
Sulfate	
Sulfite	

APPLYING THE CONCEPT: EXERCISE 5.5

Complete the following table by combining the cation and anion in a combination that represents the correct formula for the resulting compound.

Cation	Anion					
	O^{2-}	Cl^{1-}	NO_3^{1-}	SO_4^{2-}	HSO_4^{1-}	PO_4^{3-}
Na^{1+}	Na_2O					
Mg^{2+}						
Sr^{2+}						
Al^{3+}						
Fe^{2+}						
Fe^{3+}						
Zn^{2+}						
Pb^{3+}						

TERMS TO KNOW

atomic mass unit

atomic number

atomic weight

chemical periodicity

electron configuration

energy levels

families

groups

isotope

mass number

nucleus

octet rule

periods

polyatomic ion

representative elements

salts

series

transition elements

valence

CHAPTER 6

THE PROCESS OF CHEMISTRY: The Power of the Balanced Equation

GETTING FOCUSED

- *What rules do you apply to write formulas for chemical compounds?*

- *Can you determine the number of molecules or ions present in a given mass of a chemical compound?*

- *How do chemists know how much product they can make from a given amount of reactants (starting chemicals)?*

- *What is a mole and how is it used in solving problems dealing with the balanced equation?*

- *What problem-solving strategies work best for solving problems dealing with the mathematics of the balanced equation?*

- *How can you use your calculator more effectively in solving problems dealing with the mathematics of the balanced equation?*

TAKING INVENTORY

Carrying out chemical reactions to produce chemical products is the main work of industrial and research chemists. Understanding the qualitative and quantitative significance of the balanced equation is the key to success in working problems that answer such important questions as:

- How many grams of chemical A do I need to mix with chemical B to synthesize (make) 10 grams of chemical C?

- How many liters of ammonia gas can be produced if 50 liters of hydrogen are reacted with 70 liters of nitrogen gas under appropriate conditions?

- If I drop 4 grams of zinc into excess hydrochloric acid, how many milliliters of hydrogen gas can I collect?

As you can see, these questions require that you know that when certain chemicals are brought together under appropriate conditions, a reaction occurs that produces new chemical products. These questions also require that you are able to write a balanced equation that describes which substances react and which are formed and in what proportions. Just as in the previous chapters, we will begin by taking a learning inventory on the topic of chemical reactions and the balanced equation. Read the following questions and give a brief response to each. Remember, the purpose of this learning inventory is not to find a correct textbook answer to the question, but to find out what you know and believe about the topic or chemical concept addressed in each question.

Learning Inventory 3: Reactions and Equations

Consider the following reaction:

$$Zn\ (s) + 2HCl\ (aq) \rightarrow ZnCl_2\ (aq) + H_2\ (g)$$

1. Which substances are the reactants?

2. Which substances are the products?

3. Is the above equation balanced? How do you know?

4. What is the significance of the (s), (aq), and (g) in the above equation?

5. Why do you think the coefficient 2 appears in front of HCl?

6. Why do you think the subscript 2 appears below Cl?

7. How is the above equation interpreted in terms of moles?

8. If you were to put in equal amounts of zinc and hydrochloric acid in a test tube to carry out this reaction, which chemical would be used up first?

9. If you were to put in exactly 2 moles of zinc in a test tube, how many moles of hydrochloric acid would you need to completely react with all of the zinc?

10. How many moles of zinc chloride can be produced if 0.25 moles of zinc react with excess HCl?

BACKGROUND INFORMATION FOR SOLVING PROBLEMS

Before you can successfully deal with problems related to the balanced equation there are several concepts you must be able to use that relate directly to problem-solving strategies of the balanced equation. Let's review these important concepts. You will need to have your chemistry textbook available and locate each concept or topic as it is listed. Briefly summarize each topic. If it is a mathematical concept, try to develop a solution pathway to use as an algorithm for working similar problems.

Formula Writing

This topic was discussed in Chapter 4. Keep in mind that you must correctly write the formula or symbol for each reactant and product in a chemical reaction to successfully balance the equation. The following list provides some tips that you might find useful in formula writing:

- Memorize the diatomic elements. These elements do not occur naturally as single atoms. In the uncombined state when they appear as a reactant or a product in an equation, they must be written diatomically. These elements, include hydrogen (H_2), oxygen (O_2), nitrogen (N_2), fluorine (F_2), chlorine (Cl_2), bromine (Br_2), and iodine (I_2). Consider the following example:

  fluorine + hydrogen iodide → hydrogen fluoride + chlorine

Most beginning chemistry students write the equation for this reaction as follows:

$$F + HI \rightarrow HF + I$$

This is not correct because both fluorine and chlorine exist as diatomic elements in the uncombined state. The correct way to write the equation for this reaction is:

$$F_2 + 2HI \rightarrow 2HF + I_2$$

APPLYING THE CONCEPT: EXERCISE 6.1

What is wrong with the following equation?

$$H + O \rightarrow H_2O$$

- Memorize the names and formulas of the most common laboratory acids and bases. You will find that acids and bases appear quite often as reactants and products in chemical reactions. It will be to your benefit to recognize the names and be able to write the formulas for the most frequently used acids and bases. Open your textbook and look up and write the formula for the following acids and bases:

Acids	Bases
hydrochloric	sodium hydroxide
nitric	calcium hydroxide
sulfuric	aluminum hydroxide
phosphoric	ammonia
acetic	

One way to learn a list of chemical compounds and their formulas is to make a set of study "flash cards." Figure 6.1 shows how you could make a flash card to facilitate learning the common acids and bases.

front	back
4 Common acids	1. hydrochloric HCl 2. nitric HNO_3 3. sulfuric H_2SO_4 4. phosphoric H_3PO_4

Figure 6.1 A sample flash card for common acids and bases

After you complete the notecards try to think of a mnemonic device to help you remember the names of the acids. A **mnemonic device** is a way of organizing information so that the information might be recalled later. A mnemonic device that would work for learning the names of the acids would be SHNP (pronounced ship), which represents the first letter of the acids.

APPLYING THE CONCEPT: EXERCISE 6.2

Make a flash card and develop a mnemonic device for learning the names for the common laboratory bases.

- Make a copy of the polyatomic ions and become familiar with their names, formulas, and charges. Find a chart in your textbook that lists the polyatomic ions. Polyatomic ions are part of the chemical formula of many reactants and products in chemical equations. You need to have quick access to their names, formulas, and charges until you memorize the most important polyatomic ions. Photocopy the list or make a handwritten copy and paste it inside your lecture notes and laboratory notebook.

- Learn the rules for writing the formulas for ionic compounds. Remember the overall charge on an ionic compound must sum to zero.

- Learn the rules and memorize the prefixes used in naming molecular or covalent compounds. The prefixes are:

1 mono
2 di
3 tri
4 tetra
5 penta
6 hex
7 hept
8 octa

Some examples are:

CO carbon monoxide
CO_2 carbon dioxide
N_2O_5 dinitrogen pentoxide

APPLYING THE CONCEPT: EXERCISE 6.3

Make a flash card and develop a mnemonic device for learning the prefixes used in naming covalent compounds

Formula Weight

It is very important that you be able to calculate the atomic weight of an element or the formula weight of a compound. Look up the topic of formula weight in your textbook and make sure you understand the steps in calculating formula weights for compounds. Atomic weights and formula weights carry the units of amu (atomic mass units). You will discover, however, that we will use the formula weight of a compound or the atomic weight of an element expressed in grams in solving problems dealing with the mathematics of the balanced equation. An example of calculating the formula weight of a compound is given next.

Calculate the formula weight of $Ca_3(PO_4)_2$

Solution: To find the formula weight of a compound we multiply the total number of each kind of atom present in the formula times the atom's atomic weight and sum up the total.

$$3 \text{ Ca} \times 40.08 \text{ amu} = 120.24 \text{ amu}$$

$$2 \text{ P} \times 30.97 \text{ amu} = 61.94 \text{ amu}$$

$$8 \text{ O} \times 16.00 \text{ amu} = 128.00 \text{ amu}$$

$$\text{Total} = 310.18 \text{ amu}$$

Thus, the formula weight of $Ca_3(PO_4)_2$ is 310.18 amu

APPLYING THE CONCEPT: EXERCISE 6.4

Calculate the formula weight of:
1. $MgSO_4$
2. $C_6H_{12}O_6$
3. $NaHCO_3$

The Mole

This topic was discussed in Chapter 1 as one of the important concepts to build a knowledge base in chemistry. Understanding the mole concept is central to your success in working problems dealing with the mathematics of the balanced equation. It is so important that we will review it once more.

Chemical reactions take place at the microscopic or molecular level. Individual atoms, molecules, and ions react and rearrange to form new chemical products. Yet, chemists carry out chemical reactions in the laboratory at the macroscopic level. Chemists mass out chemical substances from micrograms to kilograms to convert these substances into new chemical products. Since chemical reactions occur between individual atoms, ions, or molecules, chemists needed a method of determining how many chemical particles are available to react in a given mass (grams) of a chemical substance.

Through careful research it has been proven that the formula weight of a compound, or the atomic weight of an element expressed in grams always contains a definite number of chemical particles (atoms, ions, molecules). What you need to remember is that one **mole** of an element or compound is simply the atomic weight or formula weight of the element or compound expressed in grams, and always contains 6.02×10^{23} atoms or molecules. The number of particles in a mole, 6.02×10^{23}, is known as **Avogadro's Number,** in honor of Amedeo Avogadro (1776–1856), an early Italian chemist who investigated the combining ratio of gases.

Here's how it works. The atomic weight of sodium (Na) is 23.0 amu. The mole weight of Na is 23.0 grams, and 23.0 grams of Na contain 6.02×10^{23} atoms of sodium. The formula weight of baking soda ($NaHCO_3$) is 84.0 amu. The mole weight of $NaHCO_3$ is 84.0 grams and 84.0 grams of $NaHCO_3$ contains 6.02×10^{23} formula units of baking soda. By the way, we use the term formula units of baking soda instead of molecules of baking soda because baking soda is an ionic compound. **Ionic compounds** are not composed of individual molecules, but instead are composed of positive and negative ions arranged in a crystal lattice.

Many chemical reactions involve gases as reactants, products, or both. For example, ammonia (NH_3) is prepared commercially according to the following reaction:

$$N_2\ (g) + 3H_2\ (g) \rightarrow 2NH_3\ (g)$$

Notice that both reactants and the product are all in the gaseous phase. Since it is difficult to mass out gases in grams, chemists measure out volumes of gaseous reactants in milliliters or liters to participate in chemical reaction.

Is there a relationship between volumes (liters) of gases and moles of gases? The answer is yes. Again, research has shown under certain conditions of temperature and pressure there is a certain volume of gas that contains exactly 6.02×10^{23} atoms or molecules of a gas. At a temperature of 0°C and one atmosphere (760 mm of Hg) of pressure, one mole of any gas always occupies 22.4 liters and contains 6.02×10^{23} atoms or molecules of the gas. The conditions of 0°C and one atmosphere of pressure, are known as **standard temperature** and **standard pressure (STP).** The volume of 22.4 liters measured at STP is known as the **molar volume** of gas. Remember, the molar volume (22.4 L) of a gas, regardless of its chemical formula, always contains 6.02×10^{23} atoms or molecules of that gas at STP.

So, it turns out that for gases, we use molar volumes instead of molar masses to find out how many atoms or molecules of a gas are available to participate in a chemical reaction. For example, at STP, 22.4 liters of nitrogen gas (N_2), oxygen gas (O_2), and carbon dioxide gas (CO_2) all contain 6.02×10^{23} molecules of nitrogen, oxygen, and carbon dioxide molecules respectively.

Chemists don't always use exactly one mole of a chemical substance in carrying out reactions. It is perfectly okay to use two moles or 0.25 moles of a reactant. The most important point is that if you know how many grams of an element or compound, or how many liters of a gas measured at STP is available, you know how many moles of the element or compound is present. If you know how many moles of an element or compound is present, then you know how many atoms or molecules are available to participate in a chemical reaction.

APPLYING THE CONCEPT: EXERCISE 6.5

1. What weight of Al contains 6.02×10^{23} atoms of Al?
2. What volume of NH_3 at STP would contain 0.5 moles of NH_3?
3. What is the mass of 3.01×10^{23} molecules of H_2O?

Let's stop for a moment and try to visualize the components of the mole concept by drawing a concept map. Do you remember the links that are used to show the relationship among components of a concept? If not, please review page 63 of Chapter 5.

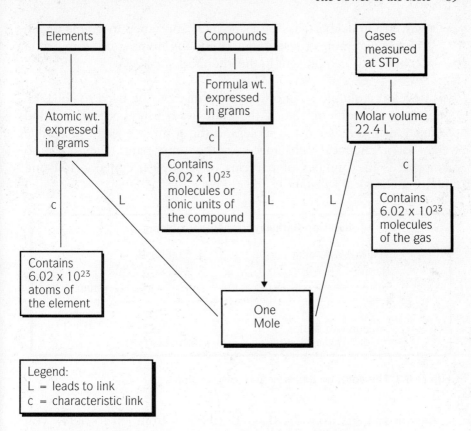

Figure 6.2 A graphic organizer for studying the mole concept

THE POWER OF THE MOLE

There are three basic types of simple mathematical problems dealing with the mole concept with which you need to become familiar. All three types of problems make use of the definition of the mole and the value of Avogadro's Number. The three mathematical concepts are:

- Given the mass of a chemical element or compound, determine the number of moles present in the chemical substance.

- Given the number of moles present in a chemical element or compound, determine the number of grams present in the chemical substance.

- Given either the mass of a chemical substance in grams or the number of moles present, determine the number of atoms, molecules, or ions, present in the chemical substance.

Let's develop a simple solution pathway that can be used to solve all three types of problems. Study the pathway below, keeping in mind the relationship between the mole and the formula weight or atomic weight of a chemical substance. Another important relationship to memorize is the number of particles (6.02×10^{23}) contained in one mole of a chemical substance.

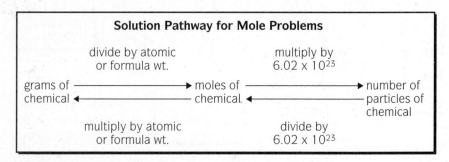

Solution Pathway for Mole Problems

divide by atomic or formula wt.	multiply by 6.02×10^{23}

grams of → moles of → number of
chemical ← chemical ← particles of chemical

multiply by atomic or formula wt.	divide by 6.02×10^{23}

Figure 6.3 The solution pathway for mole problems

As you can see from the diagram (reading from left to right), you can start with grams of a chemical, convert to moles, and then find the number of particles present in the chemical sample by following the instructions above the arrows. It is important to note that you cannot find the number of particles (atoms, ions, molecules) in a given mass of a chemical substance without first changing the sample to moles. Reading the solution pathway from right to left, you can start with a given number of particles, convert to moles, and then find the number of grams of a given chemical substance. To carry out these operations you follow the instructions below the arrows. Let's try out the solution pathway with a couple of problems.

Problem 1 How many atoms of iron are present in a 0.550 gram sample of iron?

Solution Using the solution pathway for the mole concept, we see that to find out the number of atoms present in a given mass of iron, we must first change the grams of Fe to moles of Fe. After we change the grams of Fe to moles of Fe, we next need to find the number of atoms of Fe present in the sample by multiplying by Avogadro's Number (6.02×10^{23}).

$$0.550 \; \cancel{g} \; \text{Fe} \times \frac{\cancel{\text{mole}} \; \text{Fe}}{55.85 \; \cancel{g}} \times \frac{6.02 \times 10^{23} \; \text{atoms Fe}}{\cancel{\text{mole}} \; \text{Fe}} = 5.93 \times 10^{21} \; \text{atoms Fe}$$

Study the solution to Problem 1 carefully. Note that the mathematical steps are carried out exactly as defined in the solution pathway for the mole concept. Also note that in Problem 1, we followed the solution pathway from left to right; that is, we changed grams to moles and moles to number of particles.

Problem 2 A given sample of aluminum (Al) wire contains 1.96×10^{22} atoms of Al. What is the mass in grams of the Al wire sample?

Solution Using the solution pathway for the mole concept, we see that to change a given number of Al atoms to grams we must first change the atoms of Al to moles by dividing by Avogadro's Number. Then we can find the mass of the Al wire by multiplying the number of Al moles present in the sample by the mole weight.

$$1.96 \times 10^{22} \; \text{atoms Al} \times \frac{\text{mole Al}}{6.02 \times 10^{23} \; \text{atoms Al}} \times \frac{26.98 \; \text{g Al}}{\text{mole Al}}$$

$$= 0.878 \; \text{grams of Al}$$

As you review Problem 2, note that we followed the solution pathway from right to left carrying out the mathematical instructions under the arrows. Again, it is important to note that you cannot determine the mass in grams of a number of chemical particles without first changing the particles to moles. Before we continue with the mathematics of the balanced equation, review carefully the mole concept in your textbook and make sure you can recognize the three types of mole problems given on pages 89–90 and can successfully apply the solution pathway presented for solving such problems.

APPLYING THE CONCEPT: EXERCISE 6.6

1. How many moles of CH_4 are there in 108 g of CH_4?
2. How many grams of Zn are there in 1.68 moles of Zn?
3. What is the mass in grams of 3.45×10^{14} atoms of C?

REACTIONS AND EQUATIONS: THE DOING OF CHEMISTRY

Writing balanced chemical equations is a skill that is required for all chemistry students. It is a skill that is acquired through patience and practice. Your textbook, lecture notes, and laboratory exercises are the best sources for gaining background information needed to balance chemical equations.

Beginning chemistry students often find balancing equations difficult because they can't predict what the products of a chemical reaction might be. Beginning chemistry students are not expected to predict the products of complicated equations. You will be given the names and sometimes the formula for the products of such reactions.

The best place to learn and practice balancing equations is in the laboratory. Your textbook will lay down the rules and cite examples for balancing equations, but only in the laboratory can you actually witness chemical reactions. In addition, you will be required to write balanced equations for the reactions that you carry out in the laboratory.

What information do you need to collect when you are carrying out chemical reactions in the laboratory? To write a balanced equation that describes a reaction both qualitatively and quantitatively, you need to know:

- the **Law of Conservation of Matter,** which states that during a chemical reaction, atoms are neither created nor destroyed. This is interpreted to mean that in a balanced equation the same kind and the same number of atoms must appear on both sides of the equation

- the names and formulas of the reactants

- the names and formulas of the products

In *Chemistry: An Introduction to General, Organic, and Biological Chemistry,* Karen Timberlake offers several hints for balancing equations. These suggestions are found in Table 6.1. Dr. Timberlake reminds students that these are only hints, because much of balancing equations is really trial and error.

TABLE 6.1   STEPS FOR BALANCING CHEMICAL EQUATIONS

STEP	DESCRIPTION
Step 1	Count the number of atoms for each element or ion on the reactant side and then on the product side.
Step 2	Determine which atoms need to be balanced.
Step 3	Pick one element at a time to balance. The most likely starting place is a metal, or the elements in a formula that have subscripts. Hydrogen, oxygen, and polyatomic ions are usually balanced last.
Step 4	Start balancing one of the elements by placing a coefficient in front of the formula containing that element. (Note: No changes can be made in any of the subscripts of the formulas when you balance an equation.)
Step 5	Check to see if the equation is completely balanced. Sometimes balancing one element will undo the balance of another. Then you must return to the other element and rebalance. If there are any fractions used as coefficients, multiply all the coefficients by 2 if the fraction is 1/2, and by 3 if it is 1/3. The final ratio of coefficients should be whole numbers that are not divisible by a whole number other than one.

Let's use Dr. Timberlake's steps to practice balancing a chemical equation.

Balancing Chemical Equations

Balance the following equation which shows the decomposition of potassium chlorate upon heating.

$$KClO_3(s) \xrightarrow{\text{heat}} KCl(s) + O_2(g)$$

Solution

Step 1.   Count the number of atoms of each element on both the reactant side and the product side.

Reactant Side	Product Side
1K	1K
1Cl	1CL
3O	2O

Step 2.   From the above list, it is clear that oxygen atoms need to be balanced.

Step 3.   Notice that the potassium and chlorine atoms are balanced. Oxygens are not balanced, however. There are 3 oxygen atoms on the reactant side and 2 oxygen atoms on the product side. The lowest common denominator for 3 and 2 is 6. Therefore, in order to get the same number of oxygen atoms on both sides of the equation a coefficient 2 needs to be placed in front of the reactant $KClO_3$ and a 3 coefficient in front of the product O_2.

$$2KClO_3(s) \rightarrow KCl\ (s) + 3O_2(g)$$

Steps 4 and 5.   After balancing the oxygen atoms, there are now two potassium atoms and two chlorine atoms on the reactant side but only one of each on the product side. To finish balancing the equation, place the coefficient 2 in front of the product KCL.

$$2KClO_3\ (s) \rightarrow 2KCl\ (s) + 3O_2\ (g)$$

Do a final check to make sure that the Law of Conservation of Matter is obeyed.

Reactant Side	Product Side
2 K	2 K
2 Cl	2 CL
6 O	6 O

As you might imagine, not all equations are this easy to balance. Many equations, especially a class of reactions known as oxidation-reduction reactions are sometimes difficult to balance. **Oxidation-reduction reactions** (redox) are a type of chemical reaction where electrons are transferred from one chemical substance to another. If you are required to balance oxidation-reduction equations, you will need to read and study the appropriate section of your textbook deal-ing with these equations.

APPLYING THE CONCEPT: EXERCISE 6.7

Balance the following equations:
1. $Zn + HCl \rightarrow ZnCl_2 + H_2$
2. $C_3H_8 + O_2 \rightarrow CO_2 + H_2O$

THE MATHEMATICS OF THE BALANCED EQUATION

As we have already stated, the job of the chemist is to make chemical products. Every day research and industrial chemists turn out chemical products such as agrichemicals, pharmaceuticals, plastics, explosives, dyes, synthetic fibers, and countless other chemicals. In this day and age of cost effectiveness and budget cutting, chemists must know ex-actly how much chemical reactants are needed to produce a given chemical product. They cannot afford to waste expensive chemicals needed to carry out a reaction. They must also turn out a predeter-mined amount of chemical product daily to maintain profit margins for their companies.

How do practicing chemists know how much reactant is needed and how much product can be made for a given chemical reaction? First of all, they know the balanced equation that represents the reaction that yields the product they wish to produce. Suppose you work for a com-pany that makes and sells household ammonia, a common cleaning product. To make household ammonia you bubble ammonia gas into soapy water. But first, you have to make ammonia gas. The most com-mon method of making ammonia gas is by the following reaction:

$$N_2 \text{ (g)} + 3H_2 \text{ (g)} \rightarrow 2NH_3 \text{ (g)}$$

Here is how the equation is interpreted on both a microscopic and macroscopic level:

- one molecule of N_2 reacts with three molecules of H_2 to produce two molecules of NH_3

- one volume of N_2 reacts with three volumes of H_2 to produce two volumes of NH_3

- one mole of N_2 reacts with three moles of H_2 to produce two moles of NH_3

- 28 g (2×14.0) of N_2 reacts with 6.0 g (3×2.0) of H_2 to form 34.0 g (2×17) of NH_3

- At STP, 22.4 liters of N_2 reacts with 67.2 liters (3×22.4 liters) of H_2 to produce 44.8 liters (2×22.4 liters) of NH_3

As you can see from the above statements, the balanced equation for the production of ammonia can be interpreted at the microscopic (atomic) level as well as the macroscopic (moles, grams, liters) level. Now that we know how to interpret the balanced equation quantitatively, we can return to our original question. The chemist knows how much of each reactant is needed, and how much product can be made, by carrying out simple calculations using the coefficients in front of each reactant and product as mole ratios.

A **mole ratio** is a ratio between the number of moles of any reactant or product involved in a given chemical reaction. For example, in our equation illustrating the production of ammonia, we can write the following mole ratios:

$$\frac{1 \text{ mole } N_2}{3 \text{ mole } H_2} \qquad \frac{1 \text{ mole } N_2}{2 \text{ moles } NH_3} \qquad \frac{3 \text{ moles } H_2}{2 \text{ moles } NH_3}$$

Let's see how mole ratios are used in solving problems dealing with the mathematics of the balanced equations. We will use the reaction for the production of ammonia as our example.

Mole Ratios

Ammonia is manufactured chemically under appropriate conditions of temperature and pressure by the following process:

$$N_2 \text{ (g)} + 3H_2 \text{ (g)} \rightarrow 2NH_3 \text{ (g)}$$

a. How many moles of hydrogen gas are needed to produce 100 moles of ammonia?

b. How many moles of nitrogen gas are required to react with 50 moles of hydrogen gas?

Solution a

Examining the balanced equation we notice that three moles of hydrogen gas are needed to produce two moles of ammonia gas. We will write this relationship as a mole ratio to solve the problem.

$$100 \text{ moles NH}_3 \times \frac{3 \text{ moles H}_2}{2 \text{ moles NH}_3} = 150 \text{ moles H}_2$$

Solution b

Again, the balanced equation shows that one mole of nitrogen gas reacts with three moles of hydrogen gas. As in solution a, we will use this relationship to write a mole ratio to solve the problem.

$$50 \text{ moles H}_2 \times \frac{1 \text{ mole N}_2}{3 \text{ moles H}_2} = 17 \text{ moles of N}_2$$

Three important points are worth noting in the above example illustrating the use of mole ratios in problem solving:

1. The coefficients in front of the reactants and products of the balanced equation are used to form the mole ratios used to solve the problem.

2. The problem solution only focuses on two chemical species at a time. Notice in **solution a** that we only compared the mole relationship between hydrogen and ammonia. We were not concerned at all with the reactant nitrogen gas. The same situation is true of **solution b.** Notice that we focused on the mole relationship between hydrogen gas and nitrogen gas, ignoring the ammonia gas product.

3. Note that in the solution pathway we always start with the given or stated amount of reactant or product. Next, we set up a mole ratio between the stated or given amount and the desired or unknown amount. Also note that the mole ratio is really a conversion factor. The mole ratio is set up so that the given quantity cancels out and the desired or unknown quantity and its units remain as the final answer.

Being able to identify mole relationships and setting up mole ratios between reactants and products in a balanced equation is critical in successfully solving more complicated problems dealing with the mathematics of the balanced equation.

APPLYING THE CONCEPT: EXERCISE 6.8

Consider the following reaction:

$$2N_2 \text{ (g)} + 5O_2 \text{ (g)} \rightarrow 2N_2O_5 \text{ (g)}$$

1. How many moles of N_2 are needed to combine with 200 moles of O_2?
2. How many moles of O_2 are needed to make 75 moles of N_2O_5?

PROBLEM-SOLVING STRATEGIES

Quantitative relationships between reactants and products in a balanced equation is known as **stoichiometry.** Up to this point, we have been working simple stoichiometry problems where we start with a given number of moles of a reactant or product and determine how many moles of reactant are needed, or how many moles of a product can be made. In the laboratory, however, when you are carrying out reactions, you start by massing a given quantity of a reactant in grams. It is important to remember that the quantitative relationship between reactants and products in the balanced equation is interpreted through moles not grams. If the given reactant or problem in a stoichiometry problem is expressed in grams, the mass of the given reactant or product must first be changed to moles. Likewise, if the desired amount of reactant or product is asked for in grams, the moles of the desired reactant or product must be changed from moles to grams. This should present no problem, however, because you already know how to change grams to moles and moles to grams.

Let's go back to our solution pathway developed for solving mole problems and expand it as a problem-solving model to use in working stoichiometry problems.

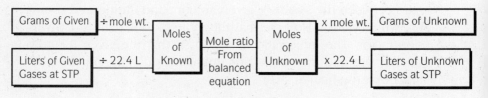

Figure 6.4 The solution pathway for stoichiometry problems

Let's apply the solution pathway given above to two examples of stoichiometry problems.

Problem 1   How many grams of hydrogen gas can be produced when 8.0 grams of aluminum reacts with an excess of hydrochloric acid?

Solution

Step 1.   We must first write a balanced equation for the reaction.

$$2Al\ (s) + 6HCL\ (aq) \rightarrow 2AlCl_3\ (aq) + 3H_2\ (g)$$

Step 2.   Identify the two chemical species of interest in the problem and set up a mole ratio between the two species.

The two chemical substances identified in the problem are Al and H_2. Using the coefficients from the balanced equation in front of Al and H_2 to set up the mole ratio, we write:

$$\frac{3 \text{ moles of } H_2}{2 \text{ moles of Al}} \quad \text{or} \quad \frac{2 \text{ moles of Al}}{3 \text{ moles of } H_2}$$

Step 3.   Using the solution pathway for stoichiometry problems as a general guideline, identify the steps needed to solve the problem. (Step 1) change grams of Al to moles of Al → (Step 2) use mole ratio of H_2 to moles of Al to determine moles of H_2 produced → (Step 3) change moles of H_2 produced to grams of H_2.

Step 4.   Convert the pathway identified in Step 3 into a series of mathematical steps making sure that unit cancellation yields the desired unit. In this case, grams of H_2.

$$8 \text{ g of Al} \times \underbrace{\frac{\text{mole of Al}}{27.0 \text{ g Al}}}_{(\text{step 1})} \times \underbrace{\frac{3 \text{ moles } H_2}{2 \text{ moles Al}}}_{(\text{step 2})} \times \underbrace{\frac{2.0 \text{ g } H_2}{\text{mole } H_2}}_{(\text{step 3})} = 0.89 \text{ g } H_2$$

Notice that in Step 2 we chose the second form of the mole ratio in order to cancel out moles of Al, leaving moles of H_2.

Problem 2

a.   How many grams of aluminum must react with hydrochloric acid to produce 4.50 liters of hydrogen gas at STP?

b.   How many atoms are contained in the mass of aluminum required to produce 4.50 liters of hydrogen gas at STP?

Solution Part A

Step 1. Write the balanced equation

$$2Al\ (s) + 6HCL\ (aq) \rightarrow 2AlCl_3\ (aq) + 3H_2\ (g)$$

Step 2. Identify the two chemical substances of interest and write mole ratio.

The two substances are Al and H_2. The mole ratio is

$$\frac{2\ moles\ Al}{3\ moles\ H_2} \quad or \quad \frac{3\ moles\ H_2}{2\ moles\ Al}$$

Step 3. Using the solution pathway for stoichiometry problems as a general guideline, identify the steps needed to solve the problem. (Step 1) change 4.50 liters of H_2 to moles of $H_2 \rightarrow$ (Step 2) use mole ratio of moles of H_2 to moles of Al to determine moles of Al needed $\rightarrow$ (Step 3) change moles of Al to grams of Al.

Step 4. Convert the pathway identified in Step 3 into a series of mathematical steps making sure that unit cancellation yields the desired unit. In this case, grams of Al.

$$4.50\ \cancel{L\ H_2} \times \underbrace{\frac{mole\ H_2}{22.4\ \cancel{L\ H_2}}}_{(Step\ 1)} \times \underbrace{\frac{2\ moles\ Al}{3\ moles\ H_2}}_{(Step\ 2)} \times \underbrace{\frac{27.0\ g\ Al}{mole\ Al}}_{(Step\ 3)} = 3.62\ g\ Al$$

Solution Part B

We know that it takes 3.62 grams of Al to produce 4.50 liters of H_2. We need to know how many atoms are contained in the 3.62 gram sample of Al. The solution pathway for this problem can be written as follows:

$$grams\ Al \rightarrow moles\ Al \rightarrow atoms\ of\ Al$$

Refer to the Solution Pathway for Mole Problems presented on page 90. Notice that this solution pathway gives all the mathematical steps needed to complete the problem.

$$3.62\ \cancel{g\ Al} \times \frac{mole\ Al}{27.0\ \cancel{g\ Al}} \times \frac{6.02 \times 10^{23}\ atoms\ Al}{\cancel{mole\ Al}} = 8.07 \times 10^{22}\ atoms\ Al$$

APPLYING THE CONCEPT: EXERCISE 6.9

Evaluate the reaction: $2Na(s) + 2H_2O \rightarrow 2NaOH(aq) + H_2(g)$ for:

1. How many grams of Na are needed to produce 6 moles of H_2?
2. How many grams of water are needed to produce 100 liters of H_2 at STP?

USING YOUR CALCULATOR

It's up to you to read critically and think carefully through the solution pathway of mole problems and stoichiometry problems. It is important for you to actually write down on paper the steps involved in the solution pathway of the problem. This is the hard part! Once you have written out the solution pathway in a series of mathematical steps and checked to make sure unit cancellation yields the desired unit you are ready to solve the problem. This is where your calculator comes in to save the day. Two important keys to learn to use in solving mole and stoichiometry problems are the "left and right parentheses" key and the exponent key.

Let's practice using the calculator in solving the problems presented in Problems 1 and 2 for Stoichiometry Problems given above. Take your calculator and locate the exponent key and the left and right parenthesis key. Now, review Step 4 in Problem 1 of the stoichiometry problems. Step 4 of Problem 1 shows the mathematical setup to the solution to the problem. Key in the following keystroke sequences: 8 times 3 times 2 divide left parenthesis 27 times 2 right parenthesis equals. These keystrokes are summarized below.

$$8 \times 3 \times 2 / (27 \times 2) = .89$$

Notice that when you use the parenthesis around (27×2) the calculator carries out the operation inside the parenthesis first, then divides the product into the product of $8 \times 3 \times 2$. Many beginning chemistry students take a chemistry problem that consists of two or more steps and solve each step individually. There is nothing wrong with this approach except that you have to key in numbers over and over again and keep a written record of each step to use as the starting point of the next step. This is not very time efficient. You need to become proficient with your calculator to maximize your time in test taking where you may only have an hour or less to complete the whole exam.

Now let's key in the keystrokes needed to solve Parts A and B of the stoichiometry problems presented in Problem 2 above. Review Step 4 for Part A of Problem 2 above. The keystroke sequence is:

4.50 times 2 times 27 divided by left parenthesis, 22.4 times 3, right parenthesis equals

The above key strokes are summarized below:

$$4.50 \times 2 \times 27 / (22.4 \times 3) = 3.62$$

Again, notice the use of the parentheses keys in the numerator of the keystroke sequence. It is very easy to carry out all three steps in the solution pathway in one entry instead of three. The end result is that you can increase your time efficiency in working homework problems and test taking so that you may have more time to read and think critically about your answers to questions on the exam.

Finally, let's review the calculator solution to Part B of Problem 2 in the stoichiometry problem given above. Review the solution to Part B, especially the series of mathematical steps comprising the solution pathway. Key in the following keystrokes:

3.62 times 6.02 exponent key 23 divided by 27 equals

The keystrokes are summarized below.

$$3.62 \times 6.02 \text{ EXP } 23 / 27 = 8.07 \times 10^{22}$$

Pressing the exponent key (EXP) after entering 6.02 places the number into scientific notation form. When you key in the numbers 23 after pressing the EXP key, the calculator interprets the number to mean 6.02 raised to the 23rd power. Remember, do not key in the number 10 after you press the EXP key. If you do this, your answer will be off by a power of ten.

TERMS TO KNOW

Avogadro's Number
ionic compounds
law of Conservation of Mass
mnemonic device
molar volume
mole

mole ratio
oxidation-reduction   reactions
standard pressure
standard temperature
stoichiometry

CHAPTER 7

WET CHEMISTRY: Solute, Solvent, and Solutions

GETTING FOCUSED

- *What are appropriate terms for describing the characteristics of solutions?*

- *What methods are used to describe the concentrations of solutions quantitatively?*

- *What problem-solving strategies work best for solving solution problems?*

- *What effect does a solute have on the physical characteristics of a solution?*

TAKING INVENTORY

As in previous chapters we will begin by taking a learning inventory about solution chemistry. Remember, the purpose of this inventory is to make an honest evaluation about your existing knowledge of concepts important in solution chemistry. Give a brief answer to each of the following questions.

Learning Inventory 4: Solution Chemistry

1. A solution is prepared by placing a given amount of solute into a given amount of a suitable solvent. How would you define the terms solute and solvent?

   _____

   _____

   _____

   _____

   _____

   _____

2. Alcohol dissolves quite readily in water, but vegetable oil separates out on the surface of water. How would you explain this observation?

   _____

   _____

   _____

   _____

   _____

   _____

3. One method frequently used in clinical chemistry to express the concentration of solutions is percent solution. For example, a 3%

solution of hydrogen peroxide might be used to clean a cut on a patient. How do you interpret a 3% solution of hydrogen peroxide?

_____

_____

_____

_____

_____

_____

4. Chemists prefer to express the concentration of solutions in terms of molarity. For example, an experiment might call for a 2.0 M (molar) solution of sulfuric acid. How do you interpret a 2.0 M solution of sulfuric acid?

_____

_____

_____

_____

_____

_____

5. If 12 grams of salt is dissolved in 600 ml of water, what is the percent concentration of the salt solution?

_____

_____

_____

_____

_____

_____

6. What is the molarity of a solution that contains 11.5 grams of aluminum sulfate $Al_2(SO_4)_3$ dissolved in 750 ml of water?

_____

_____

_____

_____

_____

CLEARING UP MISCONCEPTIONS

After completing the learning inventory, check your answers in the Appendix. How well did you do? Just as in other chapters, it is important for you to clear up any misconceptions you might have about solution chemistry. One way to clear up misconceptions is to replace incorrect knowledge and skills about a particular topic with and correct information and problem-solving skills.

As you study this chapter, open your textbook to the appropriate chapter on solutions. As you study each topic in this chapter, try to find the parallel topic in your chemistry textbook. Many topics in solution chemistry require application of quantitative problem skills. In this chapter you will be given "solution pathways" to guide you through the various types of solution problems you will find in your textbook. It is important that you develop good problem-solving skills and apply these skills to both the problems presented in your textbook and those assigned by your instructor.

THE LANGUAGE OF SOLUTION CHEMISTRY

Solution chemistry has a vocabulary that is used to describe the characteristics of solutions as well as the concentrations of solutions. It is important for you to be able to use the vocabulary of solution chemistry in developing and applying concepts of solution chemistry both qualitatively and quantitatively. Open your chemistry textbook to the corresponding chapter on solutions. The following list contains terms that are important to know and apply in the study of solutions. Find each term in your textbook. Read the supporting material describing and illustrating each term. Finally, summarize the meaning and application of each term in your own words in the spaces provided.

1. solute _____

    _____

2. solvent _____

    _____

3. unsaturated solution_____

    _____

4. saturated solution _____

    _____

5. supersaturated solution _____

    _____

6. polar solvent _____

    _____

7. nonpolar solvent_____

    _____

8. miscible _____

    _____

9. immiscible _____

    _____

10. percent concentration _____

    _____

11. molarity _____

    _____

12. molality _____

    _____

13. parts per million (ppm) _____

    _____

14. colligative properties _____

_____

15. isotonic _____

_____

16. hypotonic _____

_____

17. hypertonic _____

_____

18. osmosis _____

_____

19. osmolarity _____

_____

The above list is not inclusive. You may find that your instructor includes terms not found in this list. Your textbook may also include terms not covered in the given list. If this is the case, add these terms to your working vocabulary of solution chemistry. Remember, we have to know a given body of facts and terms about a given chemical topic before we can think conceptually or engage in problem-solving activities about the topic under study. This is why it is important to learn the terms that comprise the language of solution chemistry. These terms will be the building blocks of understanding concepts and mathematical applications to describe the complexities of solution chemistry.

APPLYING THE CONCEPT: EXERCISE 7.1

List three terms not found in your list that your textbook describes for solutions. Write a brief definition for each term.

THINKING SKILLS IMPORTANT IN SOLUTION CHEMISTRY

Comprehending the goal state of a problem and applying appropriate problem-solving strategies are crucial in studying and applying analytical concepts used in solution chemistry. You are already familiar with the concept of solution pathways for developing a framework to solve chemical problems. Developing a solution pathway for a solution to a particular problem implies that you comprehend the problem. If you truly comprehend the problem, then you can develop a solution pathway that leads to an answer to the problem.

A great deal of research has been undertaken by cognitive psychologists to determine the strategies used by good problem solvers as they develop strategies or plans to find answers to analytical problems. In *Thought and Knowledge: An Introduction to Critical Thinking*, Diane Halpern identifies the four steps important in problem solving.

Problem Translation   The first stage is comprehending the problem. This involves understanding the information that is given and recognizing the goal, or desired end state (the unknown in the problem). The problem is comprehended when you can (a) restate the information that is given and (b) specify the goal in your own words.

Problem Integration   Problem integration involves three distinct processes: representing the problem, usually as a diagram or other pictorial display, identifying the type of problem that it is, and deciding which information is relevant to finding a solution.

Solution Planning and Monitoring   Check the logic of your problem solving as you proceed. Is the answer reasonable? Is the problem set up so that unit cancellation yields the right unit needed in the final answer? Remember, there is no need to proceed with a mathematical calculation if your unit is incorrect. Check units first, then do math!

Solution Execution   In this stage, you carry out the mathematical steps called for in the solution to your problem. Using your calculator and applying appropriate key strokes is critical to this final stage of problem solving. Become familiar with your calculator. Know each key and the mathematical operation assigned to each key on your calculator. Understanding the operation of your calculator will be a tremendous asset to your success in problem solving.

As we begin to look at different types of problems involved in solution chemistry, try to keep in mind the four steps discussed above. Practice using these steps as you work out examples in your textbook, especially when you work solution problems found at the end of the chapter on solutions in your textbook.

THE MATHEMATICS OF SOLUTION CHEMISTRY

Since many chemical reactions take place when solutions are mixed together, knowing how may particles of solute are present in a given amount of solution is important. Determining the strength of a solution in terms of number of solute particles present is known as *finding the concentration of a solution*. The key to understanding and setting up solutions to concentration problems is to remember that we are always focusing on this important question:

> *How many particles of the solute in terms of grams, or moles, are present in a given volume of the solution?*

Basically, there are two ways of expressing the concentration of a solution quantitatively: percent concentration and molarity. Each method expresses the amount of solute in a given volume of total solution. We will develop problem-solving strategies for each method below.

PROBLEM-SOLVING STRATEGIES

Percent Concentration

The **percent concentration** method of expressing the concentration of solute to total volume of solution is important in health and clinical chemistry. If you look in your medicine cabinet you will probably find a bottle of isopropyl (rubbing) alcohol. Read the label and you will notice that it states "5% solution." A bottle of hydrogen peroxide is often labeled "3% solution." In hospitals, patients are infused with IV solutions that read "0.9% saline" or "5.0% glucose." How are these labels interpreted quantitatively? The key or operative word here is *percent*. We know that percent is based on part to whole. We interpret our labels as follows:

- A 5% solution of isopropyl alcohol contains 5 ml of pure isopropyl alcohol dissolved in 95 ml of water (the solvent) to make exactly 100 ml of total solution.

- A 3% solution of hydrogen peroxide contains 3 ml of pure hydrogen peroxide in 97 ml of water (the solvent) to make exactly 100 ml of total solution.

- A 0.9% solution of saline (NaCL) contains 0.9 grams of NaCl dissolved in water to give exactly 100 ml of total saline solution.

- A 5% solution of glucose ($C_6H_{12}O_6$) contains 5 grams of glucose dissolved in water to give exactly 100 ml of glucose solution.

Notice that in all of our examples, the combination of both solute and solvent yields a total volume of 100 ml of solution. It is apparent from our examples that we can express percent solutions two different ways:

- volume of solute to total volume of solution (volume/volume)

- mass of solute to total volume of solution (mass/volume)

Percent solution problems will usually fit into three types or categories. If you can identify the type or category of the percent solution problem it will help you to define and develop a solution pathway to successfully answer the problem. The three categories of percent solution problems are listed in Table 7.1.

Let's develop a general solution pathway for each of the three types of percent solution problems listed in the chart.

Type A   The solution pathway for this type of percent concentration problem is based on the definition of percent concentration:

$$\% \text{ concentration} = \frac{\text{mass of solute}}{\text{total volume of solution}} \times 100\%$$

Using the example problem for Type A given in Table 7.1, the solution pathway would be:

TABLE 7.1   TYPES OF PERCENT CONCENTRATION PROBLEMS

TYPE	EXAMPLE
A. Given the mass of solute and total volume of solution, find the percent concentration of the solution.	If 24 grams of NaCl is dissolved in 400 ml of water, what is the percent concentration of the solution?
B. Given the percent concentration of a solution and the volume of a given amount of solution, find the mass in grams of solute present.	If 250 ml of a 5% solution of glucose is administered by an IV to a patient, how many grams of glucose does the patient receive?
C. Given the percent concentration of a solution and a given mass of solute needed, determine the volume needed to provide the given mass.	What volume of a 10% solution of NaCl is needed to provide 17 grams of NaCl?

If 24 grams of NaCl is dissolved in 400 ml of water, what is the percent concentration of the solution?

Solution

$$\% \text{ concentration NaCl} = \frac{24}{400} \times 100 = 6\%$$

APPLYING THE CONCEPT: EXERCISE 7.2

What is the percent concentration of a solution that contains 8.0 grams iodine in 300 ml of water?

Type B   The solution pathway for Type B is based on using the given percent concentration of solution as a conversion factor. In Type B we are given the percent concentration of the solution and a given volume present of the solution. The unknown is the mass in grams of the solute present in the given volume of the solution. Remember a conversion factor can always be written in two forms, one form being the inverse of the other form. The percent concentration conversion factor can be written as follows:

$$\frac{\text{mass of solute}}{100 \text{ ml of solution}} \quad \text{or} \quad \frac{100 \text{ ml of solution}}{\text{mass of solute}}$$

$$\text{(form a)} \qquad\qquad \text{(form b)}$$

Since we want volume to cancel out in this type of percent concentration problem leaving an answer in grams, form a of the conversion factor is the correct form to use. The solution pathway can now be written as:

$$\text{volume of solution} \times \frac{\text{mass of solute}}{100 \text{ ml of solution}}$$

$$= \text{mass of solute in given amount volume of solution}$$

Using the example problem for Type B given in Table 7.1,

If 250 ml of a 5% solution of glucose is administered by an IV to a patient, how many grams of glucose does the patient receive?

Solution

$$250 \text{ ml} \times \frac{5.0 \text{ g glucose}}{100 \text{ ml}} = 13 \text{ g of glucose}$$

Notice that in the choosing form a of the percent concentration conversion factor, the units ml cancel out leaving the desired unit of grams.

APPLYING THE CONCEPT: EXERCISE 7.3

How many grams of NaCl are contained in 155 ml of a 2% solution of NaCl?

Type C The solution pathway for this type of percent concentration problem again uses the stated percent concentration of the solution as a conversion factor. In this case, we are trying to find a given volume of solution that contains a stated or given amount of mass. Reviewing our two forms of conversion factors shown on page 112, we see that form b will allow us to cancel out grams keeping the desired units of ml as the final unit of the problem. The solution pathway for Type C can be developed as follows:

$$\text{grams of solute given} \times \frac{100 \text{ ml of solution}}{\text{mass of solute}} = \text{ml of solution needed}$$

Using the example given for Type C in Table 7.1, we can apply the solution pathway:

What volume of a 10% solution of NaCl is needed to provide 17 grams of NaCl?

Solution

$$17 \text{ g NaCl} \times \frac{100 \text{ ml of solution}}{10 \text{ g NaCl}} = 170 \text{ ml NaCl solution}$$

APPLYING THE CONCEPT: EXERCISE 7.4

What volume of a 7% solution of glucose contains exactly 45 g of glucose?

Molarity

Molarity expresses the concentration of a solution in terms of moles of solute per liter of solution (moles/liter). This is the method preferred by chemists because it expresses concentration in terms of number of particles of solute present in a given volume of solution rather than a percentage. To calculate the molarity of a given solution we obviously need to determine two important facts about the solution:

- the total number of moles of the solute present, and

- the total volume in liters of solution resulting from the dissolving of the solute in the solvent.

Just like in percent concentration problems, there are several variations or types of molarity problems. It often helps in problem solving to categorize a problem by type and then apply a known solution pathway to the solution of the problem. Table 7.2 on pages 116–117 summarizes the different categories of molarity problems and gives a typical textbook example for each type. What you want to be able to do in solving molarity problems is to determine which type the molarity problem represents so that you can apply the appropriate solution pathway.

APPLYING THE CONCEPT: EXERCISE 7.5

Read each molarity problem and classify the problem according to type (A,B,C, or D) listed in Table 7.2. Solve each problem using the appropriate solution pathway given.

1. What mass of KCl is needed to prepare 450.0 ml of a 0.5 M solution of KCl?
2. What volume of a 1.2 M solution of NaBr contains exactly 4.00 grams of NaBr?
3. If 12.0 grams of $Ca(OH)_2$ is dissolved in 450 ml of water, what is the molarity of the solution?
4. How would you prepare 150 ml of a 0.2 M solution of HCl is you have a stock bottle of concentrated HCl which is 12 M?

COLLIGATIVE PROPERTIES OF SOLUTIONS

Before we summarize the important knowledge and problem-solving skills for colligative properties of solutions, it will be necessary for you to find and review this topic in the chapter on solutions in your textbook. Pay particular attention to the concept of **osmosis** and **osmolarity.** If you are majoring in the health sciences, the concept of osmolarity is crucial to interpreting the concentrations of fluids infused into patients intravenously. After you have reviewed the material on colligative properties of solutions and the concept of osmolarity, make sure you can summarize or carry out the calculations to provide answers to the following statements or questions:

- Explain the concept of osmosis.

- Explain and diagram the concept of osmotic pressure and osmolarity.

- Explain and diagram the movement of water across a cell membrane in which the cell is suspended in a solution that is:

   1. isotonic to the cell.

   2. hypertonic to the cell.

   3. hypotonic to the cell.

- Explain why equal molar concentrations of an ionic compound such as table salt and a molecular compound such as table sugar do not produce the same osmolarity when dissolved in water. The key to understanding this concept is that ionic compounds dissociate in water to produce two or more particles per formula units whereas molecular compounds dissolve in water in complete molecules. For example:

$$NaCl \ (s) \rightarrow Na^+ \ (aq) + Cl^- \ (aq)$$

Notice that 1 mole of sodium chloride gives 2 moles of dissolved particles.

$$C_{12}H_{22}O_{11} \ (s) \rightarrow C_{12}H_{22}O_{11} \ (aq)$$

Notice that 1 mole of sucrose (table sugar) gives only 1 mole of dissolved particles.

- Explain the relationship between the percent concentration, molarity, and osmolarity of a solution. Be able to convert mathematically, either percent concentration or molarity of a solution to its corresponding osmolarity.

THE MATHEMATICS OF OSMOLARITY

If you have read and studied the topic of osmotic pressure and osmolarity carefully in your textbook, you will have realized that when two solutions are separated by an osmotic membrane, water will move across the membrane into the solution that has the highest solute concentration. Remember, water will always flow across a membrane that exerts the higher osmotic pressure. The side of the membrane that has the highest solute concentration exerts the highest osmotic pressure.

TABLE 7.2   TYPES OF MOLARITY PROBLEMS

TYPE	EXAMPLE
A. Given the mass of the solute and the volume of the solution, calculate the molarity of the solution.	If 17.8 g of $NaHCO_3$ (baking soda) is dissolved in 450 ml of water, what is the resulting molarity of the solution?

Solution Pathway A:

$$(1) \quad 17.5g\ NaHCO_3 \times \frac{mole\ NaHCO_3}{84g\ NaHCO_3} = .208\ moles\ NaHCO_3$$

(changes grams to moles)

$$(2) \quad 450\ ml \times \frac{L}{1000\ ml} = .450\ L$$

(changes ml to liters)

$$M = \frac{.208\ moles}{.450\ L} = 0.46$$

| B. Given the molarity of the desired solution and the total volume of solution needed, find the mass in grams of solute needed to prepare the solution. | How many grams of NaOH (sodium hydroxide) are needed to prepare 3.0 L of a 0.80 M solution of NaOH? |

Solution Pathway B: It is useful to remember that in any solution, the molarity multiplied times the volume gives the number of moles of the solute present in the solution. Changing moles to grams gives the final answer called for in the problem.

moles/liter × liters = moles × grams/mole = grams of solute

0.80 moles/liter × 3.0 liters = 2.4 moles NaOH × 40 grams NaOH/mole = 96 g NaOH

| C. Given the molarity of the solution and the grams of solute needed, determine the volume of solution that contains the stated mass. | The molarity of glucose ($C_6H_{12}O_6$) in blood serum is approximately 0.28 M. What volume of blood serum contains 2.0 g of glucose? |

Solution Pathway C: The solution to this type of molarity problem requires changing the grams of solute to moles of solute and using the molarity of the solution as a conversion factor to change moles of solute to liters of solution. The molarity conversion factor can be written in two forms: (a) moles of

TYPE	EXAMPLE

solute/liter of solution or (b) liter of solution/mole or solute. We need to use the (b) form of the molarity conversion factor in order to cancel out moles and give us the required volume of solution in liters.

grams solute $\times$ grams/mole solute $\times$ Liter/mole of solute = Liters of solution

Applying the solution pathway to our example:

$$2.0 \text{ g glucose} \times \frac{\text{mole glucose}}{180 \text{ g glucose}} \times \frac{\text{liter glucose solution}}{.28 \text{ moles glucose}} = 0.040 \text{ liters glucose solution}$$

D. Dilution of concentrated solutions to less concentrated solutions. Given the molarity of a stock solution and the desired diluted molarity and volume needed, find the volume of stock solution needed to carry out the dilution.	How would you prepare 350 ml of 2.0 M solution of acetic acid if you have a stock bottle of concentrated acetic acid which is 6.0 M?

Solution Pathway D: To solve this type problem we make use of the dilution equation:

$$M_1V_1 = M_2V_2$$

Where M_1V_1 represents the molarity and volume before dilution and M_2V_2 represents the molarity and volume after dilution. When we dilute a solution the number of moles remains the same before and after dilution since we are only adding solvent. Applying the above equation to our example:

$$\frac{6.0 \text{ moles}}{1 \text{ L}} \times V_1 = \frac{2.0 \text{ moles}}{1 \text{ L}} \times .350 \text{ L}$$

$$V_1 = \frac{(2.0 \text{ moles/L})(0.350 \text{ L})}{6.0 \text{ moles/L}} = 0.12 \text{ L or } 120 \text{ ml}$$

To determine the direction of water flow across an osmotic membrane you have to determine which side of the membrane has the highest osmolarity. Quite often, this involves changing the concentration of solution given in percent or molarity to osmolarity. You need to realize, however, that you cannot change percent concentration directly to osmolarity without first changing to molarity. Let's develop solution pathways to accomplish these tasks.

Solution Pathway: Changing percent concentration to molarity

Express given percent concentration as grams per 100 ml of solution → change grams of solute to moles → change 100 ml of solution to liters → moles of solute/Liter of solution = molarity

Solution Pathway: Changing molarity to osmolarity

osmolarity = molarity × the number of dissolved particles per formula unit of solute

Example   A 0.89% solution of saline (NaCl) is isotonic to body fluids and is used frequently to infuse patients intravenously who are suffering from shock. What is the (a) molarity and (b) osmolarity of a 0.89% saline solution?

Solution a   Changing percent to molarity

0.89% saline is defined as 0.89 grams of saline/100 ml of solution.

mole weight of NaCL: Na = 23.0
$$Cl = \underline{35.5}$$
$$58.5 \text{ g/mole}$$

$$\frac{0.89 \text{ g NaCl}}{100 \text{ ml}} \times \frac{\text{mole NaCl}}{58.5 \text{ g}} \times \frac{1000 \text{ ml}}{L} = 0.15 \text{ M}$$

Solution b   Changing molarity to osmolarity

NaCl yields 2 particles (Na^+ and Cl^-) per formula unit

Osmolarity = $0.15 \times 2 = 0.30$ osmol

APPLYING THE CONCEPT: EXERCISE 7.6

A 5.0% solution of glucose ($C_6H_{12}O_6$) is isotonic to body fluids and is used as an IV in patients who are weak. What is the (a) molarity and (b) osmolarity of a 5.0% glucose solution?

TERMS TO KNOW

molarity
osmolarity
osmosis
percent concentration

CHAPTER 8

KINETICS AND EQUILIBRIUM: How Fast and How Much?

- *What factors effect how fast a chemical reaction will proceed?*
- *What are the characteristics of a reversible reaction that is in equilibrium?*
- *Can you predict how much chemical product you can make from a reversible reaction?*
- *What problem-solving skills are appropriate for solving equilibrium problems?*
- *How can you use your calculator more effectively to solve equilibrium problems?*

TAKING INVENTORY

Below are a series of questions on important concepts related to kinetics and equilibrium. The purpose of this inventory is to determine your existing knowledge about these two topics. Think about each question and write down what you know or think about the term, concept, or problem presented. Do not look up the answers to these questions. This would defeat the purpose of this exercise. Later, as you study the material in this chapter, along with the material covered in your textbook, you will have the opportunity to fill in the gaps with the appropriate content or problem-solving skill.

Learning Inventory 5: Kinetics and Equilibrium

1. What does the concept of chemical kinetics involve?

    _____

    _____

    _____

    _____

    _____

2. Name four factors that effect the rate of a chemical reaction.

    _____

    _____

    _____

    _____

    _____

3. What is a reversible reaction?

    _____

    _____

    _____

    _____

    _____

4. Many chemical reactions are reversible and eventually come to chemical equilibrium. How would you describe a reversible reaction that is in a state of equilibrium?

    _____

    _____

    _____

    _____

    _____

    _____

5. Sketch a graph that illustrates a reversible chemical reaction achieving a state of chemical equilibrium.

6. In 1888 the French chemist Henri LeChatelier developed a far-reaching generalization about the behavior of chemical systems at equilibrium. The generalization is know as LeChatelier's Principle. Below, state LeChatelier's principle in your own words.

    _____

    _____

    _____

    _____

    _____

    _____

7. Sketch the graph that illustrates a chemical reaction that is exothermic. Label the x and y axis. On the graph identify the energy of the reactants, energy of the products, the transition state, and the energy of activation.

8. Sketch the graph that illustrates an endothermic reaction. Label the x and y axis. On the graph identify the energy of the reactants, the energy of the products, the transition state, and the energy of activation.

9. The "generic equation" $aA + bB \rightleftarrows cC + dD$ represents a reversible reaction. Letters A and B represent the reactants and the letters C and D represent the products. The small letters a, b, c, d represent the coefficients needed to balance the equation. Using the generic equation, write the K_{eq} expression for this reaction.

_____

_____

_____

_____

_____

CLEARING UP MISCONCEPTIONS

If you are a beginning chemistry student, your background knowledge about kinetics and equilibrium is probably minimal. Don't be discourage if you were not very successful in the answering the questions in Learning Inventory 5. After completing this chapter, and using your textbook as a resource, your knowledge and skills related to the concepts of kinetics and equilibrium will greatly increase. Remember, the purpose of this textbook is not to provide you with all the content or mathematical background needed for a complete understanding of kinetics and equilibrium. Rather, the purpose is to provide guidelines for developing thinking and problem-solving skills. If you are in a more advanced chemistry class, you probably will be required to know the graphical interpretation and mathematical treatment of first and second order reaction rates. We will not deal with the order of reactions in this study skills text.

THE LANGUAGE OF KINETICS AND EQUILIBRIUM

To become more familiar with the concepts and mathematical treatment of kinetics and equilibrium you need to have a working knowledge of terms that relate to both topics. Using your chemistry textbook as a resource, look up the following terms. Read the supporting material related to each term and briefly provide a definition for each term in your own words.

1. kinetics_____

   _____

2. activation energy _____

   _____

3. exothermic_____

   _____

4. endothermic _____

   _____

5. molecular collision theory_____

   _____

6. activated complex_____

_____

7. reversible reaction _____

_____

8. equilibrium _____

_____

9. equilibrium constant (K_{eq})_____

_____

10. LeChatelier's Principle _____

_____

IMPORTANT CONCEPTS IN KINETICS

Molecular Collisions

Some chemical topics are better understood by using the mental strategy of visualization. **Kinetics** is the study of reaction rates, and reaction rates depend on molecular collision between reacting species. As you read and study about the **molecular collision theory** in your textbook, pay particular attention to the diagrams that illustrate the collision of reacting molecules to form products. Try to visualize molecules colliding in different orientations and with different velocities. Effective collisions are those collisions between two molecules or ions that cause bonds to break and new products to be formed. What reasons can you think of by visualization that would result in effective collisions? Similarly, what reasons can you think of by visualization that would result in ineffective collisions? Before you leave this topic in your textbook, make sure you have a good understanding of the importance of the influence of the activation energy and orientation of molecular collisions in bringing about a chemical reaction.

Energy Diagrams

Chemical reactions are classified as exothermic or endothermic reactions. **Exothermic** reactions liberate energy, usually in the form of heat as they proceed. It is important to remember that the energy content of the products formed in an exothermic reaction is always less than the energy content of the reactants. The difference in energy between the reactants and products is equal to the energy lost or given off as the reaction occurred. **Endothermic** reactions require a continuous input of energy in order to occur. Consequently, the energy content of the products of an endothermic reaction is always greater than the energy content of the reactants.

The differences between exothermic and endothermic reactions is best understood by visualizing them graphically. Locate in your textbook a diagram that illustrates the graph of an exothermic and endothermic reaction. How are the x and y axis labeled? Your textbook will identify the major points of interest on the line representing the changes in energy as the reaction progresses. Locate on both the exothermic and endothermic graphs the following points: energy of the reactants, activation energy, energy of products, and the net energy released or gained. Now, compare the two graphs in terms of the shape of the line. What differences do you see between the two graphs? Which graph is downhill? Which graph is uphill? Finally, before you leave this topic, make sure you can sketch, label, and interpret the graph of an exothermic and an endothermic reaction.

APPLYING THE CONCEPT: EXERCISE 8.1

1.   On an exothermic reaction graph, why does the energy of the products lie below the energy of the reactants on the graph?
2.   On an endothermic graph, why does the energy of the products lie above the energy of the reactants on the graph?

Factors Affecting Rates of Reactions

There are four factors that greatly influence how fast a chemical reaction will proceed toward completion. These factors are concentration, temperature, presence of a catalyst, and nature of the reactants. Using your textbook, review these factors carefully. As you study each of the above factors, make sure you understand and can discuss the underlying chemical principle that explains the effect of concentration, temperature, presence of a catalyst, and nature of the reactants on the reaction rate. Let me give you a hint! The molecular collision theory is central to explaining the effects of concentration and temperature on reaction rates. Activation energy is the key to explaining the effects of the presence of a catalyst, and the nature of the reactants on reaction rates. To monitor your comprehension of factors affecting the rates of reactions, complete Table 8.1 on page 128.

TABLE 8.1   FACTORS AFFECTING THE RATES OF REACTION

FACTOR	EXPLANATION OF HOW FACTOR AFFECTS REACTION RATES
Concentration	
Temperature	
Catalyst	
Nature of reactants	

IMPORTANT CONCEPTS IN EQUILIBRIUM

Reversible Reactions and Equilibrium

Many chemical reactions can go to 100% completion. This means that all of the reactants are converted into products. An example of such a reaction is:

$$Mg \ (s) + 2HCL \ (aq) \rightarrow MgCl_2 \ (aq) + H_2 \ (g)$$

Reactions such as this are called **nonreversible reactions.** Notice that the arrow points in only one direction, indicating that the reactants Mg and HCl are completely converted into $MgCL_2$ and H_2. There are also many reactions that can proceed forward and backward. That is, as soon as the reactants form products, the products immediately breakdown and reform products. Such reactions are called **reversible reactions.** It is important to remember that although reversible reactions can occur in two directions, one direction is usually favored. An example of a reversible reaction is:

$$N_2 \ (g) + 3H_2 \ (g) \rightleftarrows 2NH_3 \ (g)$$

Notice that in a reversible reaction, a set of double arrows is used. The double arrows indicate that the reaction can go in two directions: forwards and backwards. In this reaction, nitrogen gas and hydrogen gas react under appropriate conditions to form ammonia. As the concentration of ammonia increases, some ammonia molecules decompose and form nitrogen gas and hydrogen gas. The reaction of nitrogen gas and hydrogen gas to form ammonia is called the *forward reaction*. The decomposition of ammonia gas back down into nitrogen gas and hydrogen gas is called the *reverse reaction*. When the forward reaction rate exactly equals the reverse reaction rate, the reversible reaction is said to be in **equilibrium.** When a reversible reaction is in a state of equilibrium, the concentration of all reactants and products remains constant. At this point, you need to find the section in your textbook that deals with reversible reactions. After reading and studying the material on reversible reactions, make sure you understand the role of activation energy in determining which direction (forward or reverse) predominates at equilibrium.

The Equilibrium Constant

Chemists are interested in knowing the concentration of reactants and products in a reversible reaction. It is desirable to have a higher concentration of the products and a lesser concentration of the reactants. A useful tool to predict the relative concentration of reactants and products in a reversible reaction is the equilibrium constant. The **equilibrium constant K_{eq}** relates the concentration of the reactants and products mathematically for a given reaction. The derivation of the equilibrium equation is based on the molecular collision theory. Locate in your textbook the section that discusses the equilibrium constant, and review the mathematical development of the equilibrium constant (K_{eq}). Before applying the equation mathematically, it is most important that you be able to take any balanced reversible reaction and place it correctly into the equilibrium constant equation. You should now know the rule for accomplishing this task. Let's use the reaction for the production of ammonia to practice placing a balanced reversible reaction into the **equilibrium equation.**

$$N_2 \text{ (g)} + 3H_2 \text{ (g)} \rightleftharpoons 2NH_3 \text{ (g)}$$

$$K_{eq} = \frac{[NH_3]^2}{[N_2][H_2]^3}$$

Note that in going from the balanced equation to the equilibrium equation, we use the following information:

- The brackets around the formulas for all reactants and products refer to their equilibrium concentrations in moles/liter.

- The concentration of the products appear in the numerator of the expression and the concentration of the reactants appear in the denominator of the equilibrium expression.

- The coefficients in front of the reactants and products appear as exponents indicating the power the concentration value is to be raised.

APPLYING THE CONCEPT: EXERCISE 8.2

Write the K_{eq} expression for the following reversible reactions:
1. $N_2O_4 (g) \rightleftarrows 2NO_2$
2. $2NO_2Cl (g) \rightleftarrows 2NO_2 (g) + Cl_2 (g)$

USING YOUR CALCULATOR

It is obvious that in solving equilibrium problems you will have to raise the concentration of each reactant and product to its appropriate power as part of the solution process. The key on your calculator that allows you to carry out this function is the y^x key. Locate this key on your calculator. Let's practice using the y^x key by solving a typical equilibrium constant problem.

Problem   A sealed glass tube contains nitryl chloride in equilibrium with Cl_2 and NO_2.

$$2NO_2Cl (g) \rightleftarrows 2NO_2 (g) + Cl_2 (g)$$

At equilibrium, the concentration of the substances are:

$[NO_2Cl] = 0.00106$ M

$[NO_2] = 0.0108$ M

$[Cl_2] = 0.00538$ M

Using the above data, calculate the equilibrium constant.

Solution

$$K_{eq} = \frac{[NO_2]^2\,[Cl_2]}{[NO_2Cl]^2}$$

$$K_{eq} = \frac{[0.0108]^2\,[0.00538]}{[0.00106]^2}$$

$$K_{eq} = 0.558$$

Using the parentheses keys and the y^x key, the keystrokes are entered into the calculator as follows:

.010800 → [y^x key] → 2 → [multiply key] → .00538 → [= key] → [divide key]

.00106 → [y^x key] 2 → [= key] = 0.558493414 (round to 3 significant figures) = .558

APPLYING THE CONCEPT: EXERCISE 8.3

When the following reaction:

$$H_2\,(g) + I_2\,(g) \rightleftarrows 2HI\,(g)$$

comes to equilibrium, the concentrations are as follows:

$$[H_2] = 0.55\ M;\ [I_2] = 0.55\ M;\ [HI] = 0.39M$$

What is the K_{eq} value for this reaction?

TERMS TO KNOW

endothermic
equilibrium
equilibrium constant K_{eq}
equilibrium equation
exothermic

kinetics
molecular collision theory
nonreversible reaction
reversible reaction

CHAPTER 9

MAKING SENSE OF ACID-BASE CHEMISTRY

GETTING FOCUSED

- *What are the most important characteristics of acids and bases?*

- *What chemistry occurs when acids are placed in water?*

- *What chemistry occurs when bases are placed in water?*

- *What is the difference between weak and strong acids and bases?*

- *What is the pH scale and what does it measure?*

- *What are the best problem-solving strategies to apply when working acid-base chemistry problems?*

- *What are buffers and what biological role do they play in our bodies?*

TAKING INVENTORY

We encounter acids and bases everyday in the foods and consumer products we use in our homes. For example, grapefruits and oranges are called citrus fruits because they contain critic acid. Hair shampoos and some cleaners such as sudsy ammonia are alkaline. What distinguishes acids from bases? How do they interact with each other and with other chemicals? What biologic roles do acids and bases play in maintaining our health? As usual, we will begin out study of acid-base chemistry by taking a learning inventory. Your task is to respond to the following set of questions without using your textbook or other reference. Don't worry if you cannot answer all questions. The purpose of the learning inventory is to determine what you already know, and more importantly, to identify those topics where you need to add knowledge and problem-solving skills.

Learning Inventory 6: Acids and Bases

If a statement describes an acid, check the A column. If a statement describes a base, check the B column.

STATEMENT	A	B
Forms hydronium ion (H_3O^+) in water		
Produces solutions with a pH greater than 7		
Forms hydroxide (OH^-) ions in water		
Is a proton donor		
Is a proton acceptor		
Produces solutions with a pH less than 7		
Is an electron pair donor		
Is an electron pair acceptor		
Has a $[H_3O^+]$ greater than 1×10^{-7}		
Has a $[OH^-]$ greater than 1×10^{-7}		
Reacts with active metals to release hydrogen gas		
Reacts with hydronium ions to produce neutral water		
Found in Rolaids®, Tums®, and Alka Seltzer®		
Found in gastric juice in the stomach		

1. In chemistry, there are three types of solutions: acid solutions, base solutions, and neutral solutions. How do these three types of solutions differ from each other?

_____

_____

_____

_____

_____

2. What two chemical products are produced when equal amounts of acids and bases react with each other?

_____

3. Complete the following equation:

$$HNO_3 \ (aq) + Ca(OH)_2 \rightarrow$$

4. Three acids and their corresponding K_a values are listed below. For each, calculate their pK_a values and tell which is the strongest acid and which is the weakest acid.

ACID	K_a VALUE	PK_a VALUE	RATING
H_3PO_4	7.5×10^{-3}		
H_3BO_3	7.3×10^{-10}		
HCOOH	1.8×10^{-4}		

5. The $[H_3O^+]$ of milk of magnesia is 3.16×10^{-10}. What is the pH of milk of magnesia?

_____

CLEARING UP MISCONCEPTIONS

As you can tell from the learning inventory questions, there is a variety of qualitative and quantitative concepts that comprise the content of acids and bases. Since there is so much to learn about acids and bases, you need to have a good idea of what you already know and understand about acids and bases to focus on those concepts in which your knowledge and skill levels are weak or absent.

You need to work through this chapter with your chemistry textbook open to the chapter on acids and bases. Use your textbook as the primary source of content knowledge about acids and bases. Use this study skill book to provide you with hints for processing and remembering information about acids and bases, and to develop problem-solving strategies for acid-base problems.

THE LANGUAGE OF ACID-BASE CHEMISTRY

To gain an understanding of the chemistry of acids and bases, as well as the mathematical treatment of acids and bases, you will need to develop a working vocabulary for acids and bases. A list of terms or mathematical expressions that are important in the "language" of acids and bases is given below. Using your textbook as a reference, locate the term and study the supporting text describing or illustrating the term or expression. Finally, in your own words, write a brief description or definition of the term or expression given below.

1. Arrenhius acid _____

   _____

2. hydronium ion _____

   _____

3. hydroxide ion _____

   _____

4. Bronsted-Lowry acid _____

   _____

5. Bronsted-Lowry base _____

   _____

6. conjugate acid _____

_____

7. conjugate base _____

_____

8. strong acid _____

_____

9. strong base _____

_____

10. weak acid _____

_____

11. weak base _____

_____

12. acid solution _____

_____

13. base solution _____

_____

14. neutral solution _____

_____

15. ion product constant for water (K_w) _____

_____

16. pH scale _____

_____

17. pH formula _____

_____

18. titration _____

_____

19. indicator _____

_____

20. neutralization _____

_____

21. end point_____

_____

22. equivalent _____

_____

23. normality _____

_____

24. acid equilibrium constant (K_a) _____

_____

25. buffer _____

_____

IMPORTANT CONCEPTS IN ACID-BASE CHEMISTRY

The total concept of acid-base chemistry is linked to many important subconcepts. Each of these subconcepts is a critical link in building your knowledge base and problem solving skills in dealing with acids and bases. Although acids and bases react with other compounds, the most important chemistry they do is with each other! As you read each of the following topics, make sure you have your textbook open to the chapter on acids and bases. Remember, the purpose of this study skills book is not to teach you all of the content knowledge you need to fully understand chemical principles. Your textbook and your instructor are your primary sources of chemical content information. The purpose of this support book is to help you internalize information in a meaningful way so that you can retrieve chemical information when called upon. Another important role of this support textbook is to help you in developing solution pathways in solving acid-base quantitative problems.

Behavior of Acids in Water

By now, you should have memorized the chemical formulas of the most important acids. As you review the formulas for the acids, you will notice that they all contain the element hydrogen as part of their chemical formula. Some acids contain only one hydrogen per formula. Other acids may contain two or three hydrogens per formula unit. Certain organic acids may contain several hydrogens, some of which are connected to carbon atoms and others connected to oxygen atoms. When acids are placed in water, those hydrogens that are connected to O, Cl, F, Br, I, or polyatomic ions can be removed by water. When this process occurs, the hydrogen ion attaches to the water molecule forming a **hydronium ion.** (The hydronium ion has the formula H_3O^+.) It is important to remember that in organic acids, only those hydrogen atoms that are attached to oxygen can be removed by a water molecule. Hydrogen atoms attached to carbon atoms cannot be removed because partial charges do not exist on the carbon atom, or the hydrogen atom attached to the carbon. Locate in your textbook the section illustrating the **ionization** (removal of hydrogen) of acids when dissolved in water. Make sure you can write simple equations illustrating the ionization of acids when placed in water before you leave this section. Have you noticed that when any acid is dissolved in water you always get two products: the hydronium ion and the negative anion contained in the formula for the acid. As a way of checking your understanding of the ionization of acids in water, complete the table in Exercise 9.1.

APPLYING THE CONCEPT: EXERCISE 9.1

ACID	EQUATION ILLUSTRATING IONIZATION OF ACID IN WATER
HCl	$HCl(aq)+H_2O \rightarrow H_3O^+(aq) = Cl^-(aq)$
HI	
HNO_3	
H_2SO_4	
HCOOH	

Behavior of Bases in Water

Unlike acids, bases dissociate when placed in water. Since bases are ionic compounds, they simply split apart into positive and negative ions. As you study the section in your textbook illustrating bases dissociating in water, you will notice that water does not participate in the chemistry of the reaction. Water molecules simple "pull" the positive and negative ions away from the crystal. Consequently, in the equations illustrating dissociation of bases in water, water is not included as part of the equation. As a study check before leaving this topic, complete the table in Exercise 9.2, which illustrates dissociation of bases in water.

APPLYING THE CONCEPT: EXERCISE 9.2

BASE	EQUATION ILLUSTRATING DISSOCIATION OF BASE IN WATER
LiOH	$LiOH \ (s) \rightarrow Li^+ \ (aq) + OH^- \ (aq)$
NaOH	
$Ca(OH)_2$	
$Al(OH)_3$	

Acid-Base Models or Theories

There are three models used to describe the chemical behavior of acids and bases in chemistry. These models are:

- Arrenhius model
- Bronsted-Lowry model
- Lewis model

It is important that you are able to compare and contrast these three models. To do this, you need to use your textbook as a resource to gather background about each model. One way to organize information comparing two or more topics along several dimensions is to make use of a type of "graphic organizer" known as a matrix. A matrix organizes information so that commonalities and distinctions emerge about a particular topic. Let's develop a matrix to compare the three models for acids and bases. An example of a matrix that could be used to compare and contrast the three acid-base models is shown in Table 9.1. Use your textbook to look up the information given in the first column for each of the three models for acid-base chemistry. By filling in the empty cells, certain information about the three acid-base models can be compared. The matrix will allow you to see patterns of information that will help you to distinguish similarities and differences in the three acid-base models.

If you have successfully completed Table 9.1, you should be able to describe the three models of acid-base chemistry.

Strong and Weak Acids and Bases

Strong acids produce large numbers of hydronium ions in water solutions. **Strong bases** produce large numbers of hydroxide ions when placed in water. **Weak acids** and **weak bases** produce only a small number of hydronium ions and hydroxide ions when placed in water. You need to memorize the name and formulas of the strong acids and bases. You will find a chart listing the names of the strong acids and bases in your textbook. If you memorize the list of strong acids and bases, you will know that all other acids and bases are classified as weak.

The most important thing to remember about weak acids is that they come to equilibrium in water solutions. The reaction that takes place when a weak acid is added to water is

$$HA(aq) + H_2O \rightleftarrows H_3O^+(aq) + A^-(aq)$$

In the above equilibrium equation, HA represents the formula for any weak acid and A⁻ represents the anion released when the acid ionizes

TABLE 9.1 MATRIX FOR ACID-BASE THEORIES

	ACID-BASE MODEL		
	ARRENHIUS	BRONSTED-LOWRY	LEWIS
Acid definition			
Base definition			
Role of water			
Equation illustrating acid-base chemistry			
Role of conjugate acid and conjugate base			
Limitation of model (if any)			

in water. You already know that H_3O^+ is the acid ion that forms when a hydrogen ion unites with a water molecule. Any chemical reaction that comes to equilibrium can be placed into an equilibrium expression and an equilibrium constant can be determined for the reaction. If we let [HA] represent the concentration for any weak acid, and [H_3O^+] and [A⁻] represent the concentration of the hydronium ion and anion ion respectively, the equilibrium expression can be written as follows:

$$K_a = \frac{[H_3O^+][A^-]}{[HA]}$$

Locate in your textbook the section describing the equilibrium expressions and constants for weak acids. You will notice that the equilibrium expression for calculating the equilibrium constant (K_a) for weak acids is determined in exactly the same manner as equilibrium expressions for reversible reactions. After reviewing the section in your textbook on equilibrium expressions for weak acids, you should be able to write equilibrium expressions for any weak acid given its chemical formula. Make sure you understand why the concentration of water does not appear in the final equilibrium expression for weak acids. To check your understanding of the concept of equilibrium expressions, complete the table in Exercise 9.3.

APPLYING THE CONCEPT: EXERCISE 9.3

WEAK ACID	EQUILIBRIUM EXPRESSION FOR DETERMINING K_a
HCOOH	$K_a = \dfrac{*[H_3O^+][COO^-]}{[HCOOH]}$
CH$_3$COOH	
HF	
C$_6$H$_5$OH	

*Your textbook may use [H⁺] to represent [H_3O^+].

The numerical value of the K_a for weak acids is an indication of the acids' strength. The larger the K_a, the stronger the acid. Quite often, chemistry students are asked to rank a given list of weak acids in increasing or decreasing acid strength. The K_a values for the given weak acids can be used to accomplish this task. Working with large negative numbers, however, is somewhat cumbersome. To avoid the use of large negative numbers, chemists use another method of determining acid strength known as pK_a. When you change the K_a value of a weak acid into its corresponding pK_a value, you end up with small positive numbers. To change the K_a value of any weak acid into its corresponding pK_a value we use the following expression:

$$pK_a = -log[K_a]$$

Even though this equation might look complicated, your calculator can accomplish this task very easily. Before we use the above equation to practice finding a pK_a value for a weak acid from it's K_a value, let's review the function of the log key and the second or inverse key on your calculator. From the pK_a equation above you can see that you will need to make use of the log key on your calculator. The purpose of the negative sign in front of the log expression in the pK_a equation is to change the negative value of the log into a positive value. You will need to use the second or inverse key to carry out the reverse procedure of changing a pK_a value back into its' corresponding K_a value.

A couple of examples will illustrate the calculator keystrokes necessary to change K_a values to pK_a values and vice versa.

Example 1

Changing K_a to pK_a

The K_a for acetic acid is 1.8×10^{-5}. What is it's corresponding pK_a value?

Solution

$$pK_a = -log [K_a]$$
$$pK_a = -[log\ 1.8 \times 10^{-5}]$$

Calculator Keystrokes

[1.8] → [exponent key] → [5] → [± key] → [log key]

Your calculator window should display

-4.744727495

Multiplying the display answer by -1 and rounding to two significant figures:

$$pK_a = 4.7$$

Example 2

Changing pK_a to K_a

The pK_a of phenol is 9.9. What is the K_a value for phenol?

Solution

$$K_a = 10^{-\log pka}$$

$$K_a = 10^{-9.9}$$

Calculator Keystrokes

[9.9] → [± key] → [2nd key] → [log key]

Your calculator window should display

$$1.258925412^{-10}$$

Rounding the answer to two significant digits the final answer is:

$$K_a = 1.3 \times 10^{-10} \text{ for phenol}$$

APPLYING THE CONCEPT: EXERCISE 9.4

1. The K_a for carbonic acid (H_2CO_3) is 4.3×10^{-7}. What is the pK_a value for carbonic acid?
2. The pK_a value for nitrous acid (HNO_2) is 3.35. What is the K_a value for nitrous acid?

SELF IONIZATION OF WATER

To understand pH and pOH you need to have a good working knowledge of the ion product constant for water (K_w). You already know that acid solutions produce hydronium ions and base solutions produce hydroxide ions. What about pure water? It turns out that in pure water very small amounts of H_3O^+ and OH^- ions are present. These ions are formed when one molecule of water having sufficient energy and proper orientation collides with another water molecule. Find the equation in your textbook that illustrates the self ionization of water. Note that for every two molecules of water that collide under proper conditions, one hydronium ion and one hydroxide ion are formed. When the equation for the self ionization of water is placed into an equilibrium expression it takes on the form:

$$K = \frac{[H_3O^+][OH^-]}{[H_2O]^2}$$

Using your textbook, follow closely the explanation given for removing the concentration of water from the expression and incorporating it into the value of K to create a new constant K_w. This new expression K_w is known as the **ion product constant for water.** The new equilibrium expression for K_w can known be written as:

$$K_w = [H_3O^+][OH^-]$$

Careful measurements have shown the numerical value for K_w measured at 25°C to be 1.0×10^{-14}. The most significant implication of the K_w value for water is that there are present in the water equal concentrations of both H_3O^+ ions and OH^- ions. Since the product of hydronium ion concentration times the hydroxide concentration must always equal 1×10^{-14} in water solutions, this means that in pure water:

$$[H_3O^+] = 1 \times 10^{-7} \text{ moles/L}$$

$$[OH^-] = 1 \times 10^{-7} \text{ moles/L}$$

Here is a very important point to remember concerning the ion product constant for water. If you add acid to water, the $[H_3O^+]$ concentration goes up and the $[OH^-]$ concentration goes down. Likewise, if you add a base to water, the hydroxide ion concentration goes up and the hydronium ion concentration goes down. But the concentration of both the hydronium ion and the hydroxide ion must always equal 1×10^{-14}. The important point here is that if you know the concentration of the acid ion, you can find the concentration of the base ion using the ion product constant for water. Let's use an example to illustrate the use of the ion product constant of water to find the concentration of either the hydronium ion or the hydroxide ion in an aqueous (water) solution.

Example

Using the ion product constant of water to find acid or base concentration.

An aqueous solution of HCl is 1×10^{-5} M. What is the concentration of the OH^- in moles/L?

Solution

The $[H_3O^+]$ is given to be 1×10^{-5}. We know that in any water solution,

$$[H_3O^+][OH^-] = 1 \times 10^{-14}$$

Solving the expression for [OH⁻] we get,

$$[OH^-] = \frac{1 \times 10^{-14}}{[H_3O^+]}$$

$$[OH^-] = \frac{1 \times 10^{-14}}{1 \times 10^{-5}}$$

$$[OH^-] = 1 \times 10^{-9}$$

Calculator Keystrokes

[1] → [exp or ee key] → [14] → [± key] → [division key]

[1] → [exp or ee key] → [5] → [± key] =

APPLYING THE CONCEPT: EXERCISE 9.5

A solution of NaOH has an [OH⁻] concentration of 1.5×10^{-3}. What is the concentration of [H₃O⁺] in moles/L?

THE pH SCALE

The hydronium ion concentration is a very important characteristic of any aqueous solution. Chemists, medical researchers, doctors, and nurses frequently monitor the pH of solutions and body fluids. You already know that pK_a was developed mathematically by chemists to avoid using large negative numbers. For the same reason, chemists and health workers use a more convenient method of expressing the hydrogen ion concentration of a solution. Chemists have defined mathematically a number called **pH,** which converts the hydronium ion concentrations of acid solutions into small positive numbers. Mathematically pH is defined as,

$$pH = {}^-log[H_3O^+]$$

This is a very important equation to know and apply in solving acid base problems. You already know about the log key on your calculator and its use in solving pK_a problems. Notice the similarity of the pH equation to the pK_a equation. The keystrokes are exactly the same except you substitute in the hydronium ion concentration instead of the K_a value of the acid. You will use the pH equation along with the ion product constant expression for water to solve many types of acid-base problems.

On occasion it will be necessary to rearrange the pH equation to solve for the $[H_3O^+]$ when you are given the pH of a solution. The pH equation is then rearranged to give the following expression,

$$[H_3O^+] = 10^{-pH}$$

This equation was obtained by solving for $[H_3O^+]$ in the pH equation and taking the log of both sides of the expression.

As mentioned before, classifying problems by type helps in developing solution pathways for problems. Below are some of the most common types of pH problems with suggested solution pathways. Each type is illustrated with an example; calculator keystrokes are given where applicable.

Type 1

Given the $[H_3O+]$, find the pH of the solution.

Example

The hydrogen ion concentration of tomato juice is about 6.3×10^{-5} moles/L. What is the pH of tomato juice?

Solution Pathway

$$pH = -log[H_3O^+]$$

Solution

$$pH = -[log\ 6.3 \times 10^{-5}]$$
$$pH = -[-4.2]$$
$$pH = 4.2$$

Calculator Keystrokes

[6.3] → [exp or ee key] → [5] → [± key] → [log key]

Type 2

Given the pH of a solution, find the $[H_3O^+]$

Example

The pH of unpolluted rain is 6.2. What is the $[H_3O^+]$ of rainwater?

Solution Pathway

$$[H_3O^+] = 10^{-pH}$$

Solution

$$[H_3O^+] = 10^{-6.2}$$
$$[H_3O^+] = 6.3 \times 10^{-7}$$

Calculator Keystrokes

[6.2] → [± key] → [2nd or inverse key] → [log key]

Type 3

Given the pH of a solution, find the [OH⁻].

Example

The pH of pancreatic fluid is 8.2 What is the [OH⁻] of pancreatic fluid?

Solution Pathway

Use $[H_3O^+] = 10^{-pH}$ to find the $[H_3O^+]$. Then use the ion product constant of water to find the [OH⁻].

Solution

$$[H_3O^+] = 10^{-pH}$$
$$[H_3O^+] = 10^{-8.2}$$
$$[H_3O^+] = 6.3 \times 10^{-9}$$
$$[H_3O^+][OH^-] = 1.0 \times 10^{-14}$$
$$[6.3 \times 10^{-9}][OH^-] = 1.0 \times 10^{-14}$$
$$[OH^-] = \frac{1 \times 10^{-14}}{6.3 \times 10^{-9}}$$
$$[OH^-] = 1.6 \times 10^{-6}$$

APPLYING THE CONCEPT: EXERCISE 9.6

1. The pH of gastric juice in the stomach is about 2.2. What is the $[H_3O^+]$ concentration of gastric juice?
2. The hydrogen ion concentration of grape juice is 6.41×10^{-6}. What is the pH of grape juice?

It is important to note that only three types of solutions exist in chemistry. **Acidic solutions** contain $[H_3O^+]$ greater than 1×10^{-7}. Consequently, acidic solutions have a pH below 7. **Basic solutions** have $[OH^-]$ greater than 1×10^{-7}. Therefore, basic solutions have a pH greater than 7. **Neutral solutions** have equal concentrations of $[H_3O^+]$ and $[OH^-]$. In a neutral solution both $[H_3O^+]$ and $[OH^-]$ equal 1×10^{-7}. Neutral solutions have a pH of 7. Pure water is an example of a neutral solution.

Remember when comparing pH values of solutions that pH is a log scale. A difference of one pH unit means a multiplication by a factor of 10. For example, if you were to compare the hydronium concentration of a bar of soap, which has a pH of 8, with black coffee, which has a pH of 5, the difference in concentration would not be 3. The hydronium ion concentration in the coffee is 1000 times the hydronium ion concentration of the soap. Where did the value of 1000 come from? The difference in pH between the soap and coffee is 3 (8 - 5 = 3). Since pH is a log scale, a difference of 3 is equal to 10^3 or 1000. You will have the opportunity to measure the pH of various solutions in the laboratory using indicators and perhaps a pH meter.

NEUTRALIZATION

When acids and bases react with each other chemically they neutralize each other. This **neutralization** reaction occurs as follows:

$$H_3O^+ \text{ (aq)} + OH^- \text{ (aq)} \rightarrow 2H_2O$$

The hydronium ion from the acid reacts with the hydroxide ion from the base forming neutral water. If equivalent amounts of hydronium ions and hydroxide ions are present when the two solutions are mixed, the resulting solution contains only water, anions from the acid, and cations from the base. The products from the complete neutralization of an acid with a base is always a *salt* plus *water*. For example, when one mole of HCl completely reacts with one mole of NaOH, one mole of H_2O and one mole of NaCl is formed. This equation is illustrated below:

$$HCl \text{ (aq)} + NaOH \text{ (aq)} \rightarrow NaCl \text{ (aq)} + H_2O$$
$$\text{(acid)} \qquad \text{(base)} \qquad \text{(salt)} \qquad \text{(water)}$$

If you examine the formula for sodium chloride you will notice that the cation (positive ion), Na^+, comes from the base NaOH and the anion (negative ion), Cl^-, comes from the acid HCl. To successfully write neutralization equations for acids and bases you need to remember the following rules:

1. Neutralization reactions always produce a salt and water.

2. The formula for the salt will always contain the cation from the base and the anion from the acid.

Review carefully the section in your textbook on acid-base neutralization. To check your understanding of writing neutralization equations, complete the table in Exercise 9.7. Assume all acids and bases are in aqueous solutions.

APPLYING THE CONCEPT: EXERCISE 9.7

ACID	BASE	EQUATION ILLUSTRATING NEUTRALIZATION
HNO_3	LiOH	
HCl	$Ca(OH)_2$	
H_2SO_4	NaOH	
H_3PO_4	KOH	
HCOOH*	NaOH	

*This is the acid hydrogen that comes off. Do you remember why?

TITRATION

Titration is a laboratory technique for determining the concentration of an unknown acid or base. Almost all beginning chemistry students are required to perform an acid-base titration in the laboratory. The process of titration depends on recognizing when equal amounts (equivalents) of acid and base have been added to each other. Do you remember that when equal amounts of acid and base are present neutralization occurs? During titration a known concentration of either an acid or base is slowly added to a measured volume of an acid or base of unknown concentration. The titration ends when equal amounts of acid and base have been mixed and neutralization occurs.

What the person who is performing the titration needs is some kind of visual clue to determine when neutralization occurs during the titration. It is the job of a chemical indicator to provide the visual clue when neutralization occurs during the titration process. Chemical **indicators** are weak acids that change color at a certain pH. The indicator is *always* added to the flask containing the acid or base with the unknown concentration. You need to become familiar with the names of the most common laboratory indicators and pH range in which they experience a color change. Your textbook will more than likely illustrate several indicators, their colors, and pH ranges in which they are useful. Take a moment to locate this section in your textbook and review the role of indicators in titration.

To collect accurate data to determine the concentration of an unknown acid or base during titration, careful volume measurements must be taken. You will probably use a buret to measure the volume of the acid and base when you perform your acid-base titration in the laboratory. A buret will allow you to measure volume accurately to one tenth of a milliliter. Figure 9.1 identifies the glassware and illustrates the laboratory process of titration.

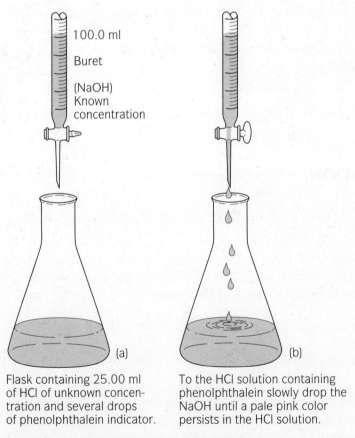

100.0 ml

Buret

(NaOH)
Known
concentration

(a)

(b)

Flask containing 25.00 ml of HCl of unknown concentration and several drops of phenolphthalein indicator.

To the HCl solution containing phenolphthalein slowly drop the NaOH until a pale pink color persists in the HCl solution.

Figure 9.1 The laboratory process of titration

THE MATHEMATICS OF TITRATION

As mentioned previously, the purpose of performing a titration is to determine the concentration of an acid or base. To determine the unknown concentration, the following activities have to be performed during the titration:

- Write a balanced equation illustrating the neutralization reaction between the acid and base used in the titration.

- Measure and record the volume of the acid or base of unknown concentration.

- Add the indicator to the acid or base of unknown concentration.

- Record the concentration (molarity or normality) and the beginning volume of the acid or base with known concentration.

- Carefully drop the acid or base of known concentration down into the acid or base of unknown concentration until the indicator changes color (neutralization occurs).

- Read carefully and record the end volume of the acid or base of known concentration.

- Record the total volume of the acid or base of known concentration used to neutralize the acid or base of unknown concentration.

After completing these steps, you now have all the numerical data needed to determine the concentration of the unknown acid or base. The mathematics used in the solution can be broken down into four steps.

1. Write the balanced equation illustrating the neutralization of the acid and base.

2. Determine how many moles of the known acid or base were used to neutralize a given volume of the unknown acid or base. To do this, we use the expression

$$\text{volume} \times \text{molarity} = \text{moles}$$
$$\text{L} \times \text{moles/L} = \text{moles}$$

3. Using the balanced equation, predict the number of moles of acid or base of unknown concentration.

4. Using the moles obtained from Step 2, and the volume of the acid or base of unknown concentration changed to liters, determine the molarity.

An example will serve to illustrate the three mathematical steps just discussed.

Example

Mathematics of acid-base titration

A chemistry student, using a 0.150 M solution of NaOH, collected the following data during the a titration experiment to determine the concentration of a sulfuric acid (H_2SO_4) sample.

volume of H_2SO_4 used	25.00 ml
beginning volume of NaOH	5.00 ml
final volume reading of NaOH	85.50 ml
volume of NaOH used	80.50 ml

Solution

Step 1. Write the balanced equation illustrating neutralization.

$$H_2SO_4 \text{ (aq)} + 2NaOH \text{ (aq)} \rightarrow Na_2SO_4 \text{ (aq)} + 2H_2O$$

Step 2. Determine the moles of NaOH used to neutralize H_2SO_4.

volume (L) × molarity (moles/L) = # of moles of NaOH

$$80.5 \text{ ml} \times \frac{L}{1000 \text{ ml}} \times \frac{0.150 \text{ moles}}{L} = 0.0121 \text{ moles NaOH}$$

Step 3. Predict number of moles of H_2SO_4 present from balanced equation.

$$0.0121 \text{ moles NaOH} \times \frac{2 \text{ moles } H_2SO_4}{1 \text{ mole NaOH}} = 0.0242 \text{ moles } H_2SO_4$$

Step 4. Determine molarity of acid.

$$25.00 \text{ ml } H_2SO_4 \times \frac{L}{1000 \text{ ml}} = 0.02500 \text{ L } H_2SO_4$$

$$\text{molarity} = \frac{0.0242 \text{ moles } H_2SO_4}{0.02500 \text{ L}} = 0.968 \text{ M}$$

APPLYING THE CONCEPT: EXERCISE 9.8

What is the molarity of a 25.00 ml sample of acetic acid (CH_3COOH) if 4.55 ml of a 1.00 M NaOH are required to neutralize the acetic acid sample? The neutralization reaction is:

$$CH_3COOH + NaOH \rightarrow NaCH_3COO + H_2O$$

BUFFERS

As you well know by now, when an acid is added to water, the pH goes down. When a base is added to water, the pH goes up. There are many instances when it is important that the pH of a solution remain relatively constant. One such instance is the pH of body fluids. The normal pH range of the blood is 7.35–7.45. There are many things in our diet and byproducts of metabolism that add considerable quantities of H_3O^+ and OH^- to the blood. Very small changes in the pH of blood can cause severe medical problems. For example, if the blood pH drops below 7.35, a condition known as acidosis occurs. A blood pH of greater than 7.45 leads to alkalosis. How does our body manage to keep the pH of the blood relatively constant? The answer of course is buffers. A **buffer** solution is a solution whose pH remains relatively constant when small amounts of H_3O^+ and OH^- are added to it. If you are majoring in the health sciences, it is important to understand the concept of buffering systems, acidosis, and alkalosis.

How are buffer solutions prepared and how do they work? This is a very important question for you to know and explain using simple equations. Locate the section in your textbook that discusses buffers and review the material carefully. Let's summarize the most important concepts concerning buffers that you should know.

Preparation of Buffers

Buffering solutions are prepared by adding to water equal quantities of a weak acid and its conjugate base. The conjugate base of an acid is the anion that results when the hydrogen is removed from the acid by water or some other base. For example, acetic acid (CH_3COOH) is a weak acid. Its conjugate base is CH_3COO^-, the acetate ion. Notice that the only difference between acetic acid and the acetate ion is a hydrogen atom. Suppose you wanted to prepare a buffer solution of acetic acid and its conjugate base, the acetate ion. Since acetic acid is a weak acid, it doesn't produce very many acetate ions or hydronium ions. Because of this, you need another source to supply additional acetate ions. For example, you could use the salt sodium acetate ($NaCH_3COOH$) as a source of acetate ions. Sodium acetate is an ionic compound and dissolves readily in water, supplying an abundance of CH_3COO^- ions.

How Buffer Solutions Work

To maintain constant pH, a buffering system must have a mechanism for soaking up hydronium and hydroxide ions when they are added to the solution. It is the job of the conjugate base of the weak acid to soak up hydronium ions. It is the job of the weak acid to soak up hydroxide ions. We can illustrate how the acetic acid/acetate ion buffering system works by use of the following equations:

Role of Weak Acid

$$CH_3COOH + OH^- \rightarrow CH_3COO^- + H_2O$$
(removes base ions from solution)

Role of Conjugate Base

$$CH_3COO^- + H_3O^+ \rightarrow CH_3COOH + H_2O$$
(removes acid ions from solution)

One final note before we leave the topic of how a buffering system works. Every buffering system has its own unique pH. How can you predict what the pH of a buffering system will be? It turns out that the pH of the buffering system is exactly equal to the pK_a of the weak acid used to prepare the buffering system. In your textbook you will find a chart that list the weak acids, their K_a values, and their pK_a values. Suppose you are asked to predict the pH of a buffering solution that is prepared by adding equal molar concentrations of formic acid (HCOOH) and the salt sodium formate (NaHCOO). To find the answer to this question, you would locate the chart in your textbook giving the pK_a values for the weak acids. Go down the list until you locate formic acid. Notice that the pK_a value for formic acid is 3.75. Therefore, the pH of a formic acid/formate ion buffering system is 3.75. Don't worry if your textbook only gives you the K_a value for formic acid. It is very easy to convert a K_a value for any weak acid to its corresponding pK_a value. We discussed this at the beginning of this chapter. You may want to review the section of this chapter that shows the mathematical relationship between pK_a and pH if you have forgotten the relationship between pK_a and pH for a weak acid.

Blood Buffering Systems

There are three buffering systems used in the body to maintain the pH of the blood between 7.35 and 7.45. These three systems are the H_2CO_3/HCO_3^- system, the $H_2PO_4^-/HPO_4^{2-}$ system, and the blood proteins. Of these three buffering systems, the carbonate system is the most important. Make sure you take the time to read and study these three buffering systems in your textbook and be prepared to discuss their mechanisms (how they work). It is important that you be able to use equations to illustrate the ability of the first two buffering systems to soak up acid ions and base ions to maintain a stable pH environment.

TERMS TO KNOW

acid solution	neutralization
base solution	neutral solution
buffer	pH scale
hydronium ion	strong acid
hydroxide ion	strong base
indicator	titration
ion product constant	weak acid
for water	weak base
ionization	

CHAPTER 10

UNDERSTANDING CHEMISTRY BY DOING CHEMISTRY: The Laboratory Experience

GETTING FOCUSED

- *Do you feel "comfortable" working in the chemistry laboratory?*

- *Do you know where all safety items are located in the laboratory and how to use them?*

- *What do you do to prepare for your laboratory experiment each week?*

- *What can you do to get the most out of each laboratory experience?*

- *What types of information should be incorporated into a "good" laboratory report?*

THE ROLE OF THE CHEMISTRY LABORATORY

It is informative and interesting to learn the principles of chemistry by reading your textbook and listening to your instructor explain and illustrate chemical concepts. But it is exciting to see chemicals react, solutions turn color, precipitates form, and to recover and identify the products of chemical reactions. This is the "doing" part of chemistry—the part where chemical concepts are verified, where mathematical equations presented on paper become tools to identify physical and chemical properties of elements and compounds. This is the role of the chemistry laboratory: to provide a link between the content you learn from print and lectures to the process of discovery that defines scientific inquiry.

Many students enter the chemistry laboratory with a sense of apprehension, uncertainty, and sometimes even fear of the unknown. Most of this apprehension and concern is because of the lack of experience of working in the chemistry laboratory. Some common concerns that students often voice include:

"Who writes these lab books? Am I suppose to understand what to do in lab when I read over the instructions?"

"My lab instructor makes me organize and write down everything I do in lab in a lab notebook. Then she makes me write a lab report. Why do I have to write the same thing down twice?"

"The sink at my lab station has six nozzles, what in the world are they all used for?

"My lab instructor gets upset when I ask for a "whatcha callit" or a "chemistry gadget." Surely, he doesn't expect me to learn the names of all this stuff!"

"I'm afraid to light a bunsen burner. It might explode!"

"I read the lab and studied the diagrams, but I still don't know how to set up the glassware and connections for this lab!"

"I always spill solutions when I try to transfer from one piece of glassware to another. Isn't there a set of rules I should know to do this correctly?"

"Every chemical in this lab either smells bad, burns your skin, or turns your fingers purple or yellow."

Do you see yourself in any of these quotes? The chemistry laboratory should be a place of interest, excitement, and yes, even enjoyment! However, making the most out of your laboratory experience requires that you be willing to adequately prepare for each laboratory session. Being prepared will allow you to enter the laboratory focused, organized, and knowledgeable about the chemical concepts, laboratory apparatus, and experimental procedures needed to complete the laboratory experiment.

ORIENTATION TO THE LABORATORY

It is important that you be familiar with the physical layout of the chemistry laboratory. There are several safety items in every chemistry laboratory that you should locate and know how to operate. These safety items include fire blankets, fire extinguisher, safety showers, eyewash fountains, first aid kits, and other emergency equipment. It is also a good idea to locate and know how to use the master cutoff valves for both gas and water. Your laboratory instructor will more than likely point out and demonstrate the use of all safety and emergency equipment found in the laboratory. If by chance your laboratory instructor does not do this, don't hesitate to ask about the location and use of safely equipment in your laboratory.

You will probably be assigned a permanent laboratory station in which your laboratory glassware will be stored. Take the time to study the layout of the gas and water valves at your station. You might be a bit confused by all of the different nozzles branching off the hot and cold water taps. Ask your laboratory instructor to explain the function of all the attachments. Learn which nozzle is the aspirator. You will use this attachment frequently in the laboratory to carry out vacuum filtration. Students have connected their bunsen burners to the aspirators. Knowing the layout of the water and gas system at your laboratory station will save you from making such mistakes.

Not all the equipment you will be using will be housed in your laboratory station drawer. For example, you may be using centrifuges, spectrophotometers, and pH meters during the course of your work. Equipment of this type is usually located in a specific section of the laboratory. You can manage your time more efficiently by knowing where to find equipment, chemical reagents, and other supplies needed to carry out your assigned laboratory investigation.

PREPARING FOR THE LABORATORY EXPERIMENT

Preparing for the laboratory exercise is the key to a successful experience. The first step is reading and studying the assigned laboratory experiment. If you have read and studied the assigned exercise, then you should be able to:

- State the purpose of the laboratory.

- Write and outline the experimental procedure in your own words in your laboratory notebook.

- Be aware of any safety concerns or procedures to be followed during the course of the laboratory exercise.

- Know how to dispose of any waste chemicals or materials generated during the course of the laboratory exercise.

- Identify all glassware and equipment to be used in the laboratory exercise.

- Sketch the setup of connections of glassware or other equipment needed to carry out the laboratory exercise.

- Answer all pre-laboratory questions related to the assigned laboratory.

- Identify and know the use of all mathematical equations needed to complete analysis of data collected during the course of the laboratory exercise.

- Construct all data tables needed to record data during the course of the laboratory experiment.

More than likely, you will be required to attend a pre-laboratory lecture before going to the laboratory. The purpose of this pre-laboratory lecture is to make sure that you understand procedures, safety precautions, and mathematical applications needed to complete the exercise. It is during the pre-laboratory lecture that your instructor will demonstrate the proper use of any new equipment you have not previously seen or used before. Quite frequently, part or sections of the assigned laboratory may be omitted or modified by your instructor. Make sure you note any changes in procedure or chemical reagents identified by your instructor. Write down the name and formula for any substituted reagents (chemicals) so that you will recognize the chemical reagents in the laboratory when they are needed.

This is the time for you to ask any questions about any aspect of the laboratory that you do not understand. Pre-laboratory lectures usually last from thirty minutes to an hour. Even though you might be anxious to get into laboratory, attending the pre-laboratory lecture is time well spent. Attending this lecture will give you a chance to clear up any concerns you may have about procedures, data collection, mathematical applications, and equipment use.

HOW TO GET THE MOST OUT OF EACH LABORATORY EXPERIENCE

The major activity you should be involved in during the activity is "focusing" on the chemistry you are performing and the measurements you are taking, not desperately trying to find out what to do next. To keep focused, you have to feel confident about your instrumentation skills, observation skills, analytical skills, and measurement skills. How do you gain this sense of confidence in the laboratory? You gain this confidence by knowing for each laboratory:

- the purpose of the lab
- what chemical concepts are under investigation
- what measurement instruments to use
- what measurement to take
- what observations to make
- what data to collect
- how to organize and interpret the data
- how to interpret the data graphically if applicable
- how to summarize the data

Gaining Confidence in Measurement

Many chemistry students have a great deal of difficulty making and recording measurements in the laboratory. Take the time before you begin each laboratory to examine each measuring device you will be using. What is the name of the instrument? How is it calibrated? What is the smallest unit of measurement the instrument will allow you to record? How many significant figures can you record using the instrument? If you cannot answer these questions after examining the measuring device, ask your instructor to explain the use and calibration of the instrument. Being able to answer these simple questions will greatly increase your measurement skills and help to improve your laboratory grade.

Gaining Confidence in Observing

Another important skill in the laboratory is that of observation. All of us by nature are poor observers of what goes on around us. But observing chemical phenomena in the laboratory is central to understanding the chemical concepts related to the laboratory investigation. A solution changes color when another solution is added to it. Why does this occur? Heat is evolved when two solutions are mixed? Why? When two solutions are mixed, an insoluble precipitate settles on the bottom. What chemical principle is related to this phenomena? When zinc is placed in hydrochloric acid, hydrogen gas forms; but when copper is placed in the same acid, no hydrogen gas forms. How do you explain this?

Train yourself to become a good observer in the laboratory. Take the time to record your observations in your laboratory notebook. Make notes about what you see and try to relate your observations to chemical concepts you have learned in lecture or read in your textbook. You will find that good written observations will make the task of writing your laboratory report much easier. Becoming a good observer in the laboratory will help you tremendously in building your knowledge of chemical principles and concepts.

Gaining Confidence in Data Collecting

Data is all the information you collect and record in your laboratory notebook during the course of the laboratory experiment. There are two types of data. Qualitative data is descriptive data. Some examples of qualitative data are:

- The solution turned red when sodium hydroxide was added to it.
- The strip of zinc turned black when placed in a solution of copper sulfate.
- Bubbles of gas formed in the test tube containing HCl when a piece of Al wire was placed in the test tube.

Quantitative data is numerical data. Examples of quantitative data include:

- The mass of the copper metal is 22.67 grams.

- The amount of copper oxide recovered is 5.21 grams.

- The temperature of the solution is 34.6°C.

The most important thing to remember about quantitative data that you collect in the laboratory is that there is a limit to the number of digits you can record when you take a measurement. The number of digits you can record is related to the calibration of the instrument used to make the measurement. This is why you must take the time to study each measuring device and identify the smallest unit of measurement. For example, using a graduated cylinder you can measure the volume of a liquid to the nearest milliliter (mL). But, if you use a buret, you can measure the volume of a liquid to the nearest 0.1 mL. Most chemistry laboratory instructors are very picky about students recording the proper number of digits when making measurements. All the digits in a measurement known for certain, plus the first digit you have to estimate are known as significant digits. So what it boils down to is that you must know the limitation of your measuring instrument and record only significant digits in your data tables. If you attend to this task in a consistent manner you can greatly improve your lab report grades.

PLANNING AND WRITING A "GOOD" LAB REPORT

The Laboratory Notebook

More than likely, your laboratory instructor will require you to keep a laboratory notebook. Do not confuse the laboratory notebook with the laboratory manual. The laboratory notebook is a permanent record of your qualitative and quantitative observations made during the laboratory. Your instructor may require that you record all information in your notebook in ink. If you make a mistake during recording, simple draw a line through the incorrect information and replace it with the correct information. There are a few things you need to attend to in your laboratory notebook each week before going to your scheduled laboratory session. You will need to:

- Briefly summarize the laboratory procedure so that you will have a good understanding of what to do during the laboratory period.

- Write down the purpose of the laboratory experiment.

- Identify all equipment to be used during the laboratory, and make a sketch showing setup of glassware, if applicable.

- Design and construct data tables that will facilitate the collecting and organizing of qualitative and quantitative data collected during the laboratory.

You might ask, "Why do I need to keep a lab notebook when everything I need to know is in the laboratory manual? For one reason, it helps you in getting prepared for the laboratory. By attending to each of the above items each week before coming to laboratory, you have increased your knowledge and confidence in carrying out the assigned experiment. Another good reason for keeping a laboratory notebook is that this is the way "real scientists" keep records of their experimental projects. The chemistry laboratory should be a place of discovery, and you should get a "feel" of how chemists go about doing science.

The Laboratory Report

Each week you will be required to hand in a laboratory report. The laboratory report is written based upon the findings recorded in your laboratory notebook. Your laboratory instructor will give you specific instructions about the format of the laboratory report. Most laboratory reports contain the following components:

- Introduction

- Procedure

- Tabulated experimental data

- Calculations related to data

- Discussion of results and comparison to known values when available

- Conclusion

More than likely the majority of your laboratory grade will come from your performance on your reports. Consequently, you want to do a good job in writing them. What can you do to ensure the best grade possible on your laboratory reports? First of all, make sure you follow the format given to you by your instructor for writing your laboratory report. Don't take "short cuts." If you leave out a section, if you don't show calculations, if you don't construct and attach graphs if called for, you can be assured that you are going to be penalized.

Make an effort to hand in a neat, organized laboratory report. Nothing is more frustrating to an instructor grading your report than trying to "mentally wade through" a messy laboratory report to locate information. If you do not have a word processor at home, use the computer laboratory at your school. A report written on a word processor looks professional, is easy to read, and shows that you care about the quality of the work you want your instructor to evaluate. If you must handwrite your laboratory report, make every effort to write legibly and neatly.

Pay attention to the numerical data you collect. Make sure you record the correct number of digits for each measurement. If you add, subtract, multiply, or divide measured numbers collected during the laboratory, make sure you round to the appropriate number of significant digits before you record your answers. If you put down every decimal point displayed on your calculator your instructor is going to lower your laboratory grade. Many students, even though they are repeatedly warned, lose points on their laboratory reports because of not practicing significant digits. Don't let this happen to you! Think about each measurement you take, or treat mathematically, and record the correct number of digits.

Show all calculations requested in your laboratory manual in your report. Using a calculator to get the answer doesn't relieve you of the responsibility of showing the solution setup of the calculation, including unit cancellation. By showing the solution pathway to all calculations requested in your laboratory manual you are demonstrating that you understand the mathematical application needed to solve a chemical problem.

Do a good job of analyzing the data and discussing the relationships or trends in the data. Many lab exercises require that certain data be graphed. If graphs are to be included in your report, make sure you identify the variables on the x and y axis and their units. Give the graph an appropriate title. Calibrate your graph so that the shape of the line covers most of the graph. Identify the slope of the graph and tell what it represents.

Finally, summarize the main points of the laboratory in your conclusion. Did you accomplish the purpose of the lab? Do your experimental results support or verify the theory, law, or chemical principle under investigation? What were your sources of error? Did these errors impact on your experimental results?

Writing a good laboratory report takes time, patience, and practice. But the rewards are well worth the effort. You learn a lot of chemistry in the process. And you can take pride in handing in a document that reflects your ability to collect, organize, analyze, and evaluate data that explains or verifies chemical principles you read about in your textbook, or hear about in class.

GLOSSARY

acid dissociation constant (K_a) the equilibrium constant for the dissociation of a weak acid. If we let HA be the general equation for a weak acid, the expression is equal to:

$$K_a = \frac{[H^+]\,[A^-]}{[HA]}$$

atomic mass units (amu) a system of measurement for describing the masses of the atoms of the elements. One amu is defined as one-twelfth of the mass of a C-12 atom. C-12 is one of the isotopic forms of carbon.

Avogadro's number the number of particles in one mole: 6.02×10^{23}.

buffer a solution containing a weak acid and its anion. The anion is supplied by a salt that contains the same anion as the weak acid. The buffer system acts to prevent a large change in the pH of a solution.

calorie a measurement of heat energy. One calorie is the amount of heat needed to raise the temperature of one gram of water 1 degree Celsius. 1000 calories is equal to 1 kilocalorie.

colligative properties a property of a solution that depends upon the number of particles dissolved in the solution, not what type of particle. Some important colligative properties of solution are boiling point elevation, freezing point depression, vapor pressure lowering, and osmotic pressure.

conversion factor a ratio of two forms of a unit that are equivalent. Conversion factors are used to convert one unit to another. For example, converting grams to milligrams, or liters to quarts.

covalent bond a bond between nonmetallic elements in which electrons are shared.

density the ratio of the mass/volume of a substance. Density is constant for a pure substance and can be used to identify the substance. The formula for density is D=M/V.

167

endothermic reaction   a chemical reaction that absorbs energy as the reactants are converted into products. The products have more energy than the original reactants.

energy level or shell   region around the nucleus of an atom in which there is a high probability of finding electrons. The first energy level can hold only two electrons. The second energy level can hold eight electrons. Atoms with large numbers of electrons have many energy levels.

equilibrium constant (K_{eq})   expression showing the concentrations of reactants and products in reaction at equilibrium. The equilibrium expression constant for the equation

$$aA + bB \rightleftharpoons cC + dD,$$
has the form:
$$K_{eq} = \frac{[C]^c[D]^d}{[A]^a[B]^b}$$

exothermic reaction   a chemical reaction that releases energy as the reactants are converted into products. The products have less energy than the original reactants.

graphic organizer   a way of showing visually how concepts are linked together. Concept maps or graphic organizers are a learning tool to help show and learn the connections or components that comprise concepts.

heterogeneous substances   substances, such as mixtures (air and blood), whose composition does vary or is not uniform throughout.

higher-order thinking skills (HOTS)   skills used in learning and solving problems that require the learner to go beyond memorization. Some examples include comprehension, application, analysis, synthesis, and evaluation.

homogeneous substances   substances, such as elements and compounds, that have the same composition throughout. That is, their composition does not vary.

ion   a charged atom. An atom that loses electrons is positively charged. An atom that gains electrons is negatively charged.

ion product constant (K_w)   the product of H_3O^+ and OH^- molar concentrations in water or any aqueous solution. The expression and value is:

$$K_w = [H_3O^+][OH^-] = 1 \times 10^{-14}$$

ionic bond   bond between metallic and nonmetallic elements in which electrons are transferred, resulting in positive and negative ions being attracted to each other.

isotopes   forms of an atom that vary in mass. Isotopes of an atom have the same number of protons and the same number of electrons. They differ in the number of neutrons.

joule   a measurement of heat energy and all other forms of energy. 4.184 joules = 1 calorie.

kinetic energy   energy of motion. The amount of kinetic energy of a chemical substance depends on the mass of the particles of the chemical substance and how fast the particles are moving (i.e., velocity). The formula for determining the kinetic energy of a substance is

$$KE = \frac{MV^2}{2}$$

kinetics   that part of chemistry that deals with rate of chemical reactions and the factors that affect rates of reactions.

LeChatelier's Principle   when a stress is applied to a system at equilibrium, the equilibrium shifts so that the stress can be relieved. When a reversible reaction comes to equilibrium, certain factors such as changes in temperature and pressure (stress factors) upset the equilibrium causing a shift to relieve the stress.

molar volume   the molar volume of a gas is 22.4 liters measured at STP.

molarity   a method of expressing the concentration of a solution in terms of moles of solute per liter of solution.

mole   the atomic weight of an element expressed in grams, or the formula weight of a compound expressed in grams, is equal to one mole of the element or compound. One mole of an element or compound always contains 6.02×10^{23} atoms or molecules.

mole ratio   a ratio between the number of moles of any reactant or product in a given reaction.

neutralization   a chemical reaction in which the hydronium ion from an aqueous solution of an acid reacts with a hydroxide ion from an aqueous solution of a base to produce water.

$$H^+ \text{ (from acid) } + OH^- \text{ (from base) } \Rightarrow H_2O$$

octet rule   noble gases all have eight electrons in their outermost energy shell. This configuration gives the noble gases high stability. Atoms that do not have eight electrons in their outermost energy shell seek other atoms to lose, gain, or share electrons in order to achieve an octet of electrons.

percent concentration   a method of expressing the concentration of a solution in terms of grams of solute per 100 ml of solution.

pH   a number, usually between the range of 0 and 14, that describes the acidity of an aqueous solution.

$$pH = -log[H_3O^+]$$

pK$_a$ of a weak acid   the negative log of the K$_a$ for a weak acid. The formula is:

$$pK_a = -log[K_a]$$

The pK$_a$ of a weak acid is a small positive number and is easier to use and interpret than the corresponding K$_a$ value for a weak acid.

polyatomic ions   a group of atoms that are covalently bonded to each other but bear an overall charge. Polyatomic ions usually remain intact during chemical reactions. Some examples of polyatomic ions are CO_3^{2-} and NH_4^+.

potential energy   energy that is stored. The most important type of chemical potential energy is the energy that is stored within the chemical bonds that hold ions and molecules together in compounds.

reversible reaction   a reaction that can proceed in either the forward or the reverse direction. As soon as some of the reactants are converted into products, the products decompose and reform the original reactants.

scientific notation   a method of expressing large and small numbers in science and mathematics. Numbers placed in scientific notation have the form $M \times 10^n$, where M stand for any positive number between 1 and 10, and n stands for the power of 10 to which you must raise M to obtain the number. For example, 106,000 expressed in scientific notation is written 1.06×10^5.

significant digit   all the digits in a measurement that are known for certain, plus the first digit you have to estimate. The number of significant digits you record when taking a measurement depends on the measuring instrument you are using.

solution   a homogeneous mixture composed of a solute and a solvent.

solution pathway   a series of mathematical steps that provides a mathematical solution to a chemistry problem.

specific gravity   a ratio of the density of a pure substance compared to the density of water. Specific gravity has no units.

specific heat   the amount of heat (in calories or joules) necessary to raise the temperature of one gram of a substance by one degree Celsius. Metals have very low specific heats. Water has a high specific heat; 1.00 cal/g degree.

STP   standard temperature and pressure. Standard pressure is 1 atmosphere, 760 mm Hg, 760 torr. Standard temperature is 0 degrees Celsius or 273 Kelvin.

temperature   a measure of the average kinetic energy of the particles of a substance. When you take a temperature reading of a liquid in the lab using a thermometer, you are measuring the average kinetic energy of the molecules of the liquid.

titration   a laboratory technique in which the concentration of an unknown acid or base is determined by using a known concentration of an acid or base.

APPENDIX: ANSWERS TO QUESTIONS

CHAPTER 1

Exercise 1.1

1. Knowledge or recall

2. Knowledge, comprehension, and application. In carrying out a titration you would have to know the characteristics of acids and bases as well as what an indicator is and the role it plays in the titration. Finally, you would use the application level of thinking in applying the principle of acid/base neutralization as the basis for carrying out the titration.

3. Comprehension. When you write a summary of an article, you are translating the information and generalizing in your own words. You are demonstrating that you comprehend the material being discussed.

4. Comprehension and knowledge. If you recognize the independent, dependent, and controlling variables in an experiment, you are demonstrating that you know what these variables are (knowledge) and can interpret relationships (comprehension) between variables.

5. Application. In solving chemistry problems, you are constructing solution pathways or using equations to solve problems.

6. Analysis, comprehension, and application. Here, you have to use multiple higher-order thinking skills. You have to comprehend the problem, construct a solution pathway, and explain the solution to your classmates.

7. Knowledge. You are simply recalling facts.

8. Synthesis, knowledge, comprehension, application, and analysis. When you design a laboratory experiment you have to use synthesis plus all the higher-order thinking skills below this level.

You must have a knowledge about ions and their characteristics. You must comprehend the process of solubility and ionic reactions. You must apply the lab techniques of filtration, centrifugation, or distillation. You must analyze reactions and products produced, and, finally, you must compile all the data in order to complete the project.

9. Comprehension plus knowledge. You must know certain facts or characteristics about ionic bonds and covalent bonds in order to discuss these types of bonds. You use these facts to interpret or explain the differences between ionic and covalent bonding.

10. Evaluation plus other thinking skills below it. When you are making judgments or critiques based on certain criteria or evidences, you are using the evaluation level of higher-order thinking skills.

CHAPTER 2

Exercise 2.1
Answers will vary.

Exercise 2.2
Answers will vary.

Exercise 2.3
Answers will vary.

Exercise 2.4
Pressure is the independent variable, temperature is the dependent variable, and volume is the controlling variable.

Exercise 2.5
1. 9.1 liters
2. 2.19 meters
3. 0.855 pounds

CHAPTER 3

Exercise 3.1

0.540 kg (hint: convert all three to the same unit and compare)

Exercise 3.2

1. 5

2. 3

3. 5

Exercise 3.3

Atoms of an element are identical in that they contain the same number of protons and electrons. Since most elements have at least two isotopic forms, they vary in mass because of a different number of neutrons in the nucleus.

Exercise 3.4

1. ionic

2. covalent

3. covalent

4. ionic

5. covalent

Exercise 3.5

b.   18 grams of water

Exercise 3.6

The coefficients 2, 5, and 2 represent the number of moles of the N_2 and O_2 combining and the number of moles of N_2O_5 formed.

CHAPTER 4

Learning Inventory 1

STATEMENT	TRUE	FALSE
1. If you use the right instrument and measure carefully, your measurement will be free of errors.		✓
2. In the metric system, mass is measured in grams, length is measured in meters, and volume is measured in liters.	✓	
3. All measurements have a limit to the number of digits you can record to describe that measurement.	✓	
4. All measurements consists of two parts: a number and a unit.	✓	
5. Accuracy and precision in measurement really mean the same thing.		✓
6. When you add, subtract, divide, and multiply measured numbers using a calculator, the correct answer is the answer displayed on the answer window of the calculator.		✓
7. Scientific notation is a method of expressing extremely large and small measured numbers.	✓	
8. Measured numbers and their units are not treated the same in mathematical calculations.		✓
9. For all practical purposes, mass and weight have the same meaning.		✓
10. Exact numbers have no uncertainty in measurement, whereas measured numbers always have some uncertainty.	✓	

Exercise 4.1

1. 6.78×10^{-8}, 1.72×10^{4}, 1.244×10^{6}, 8.091×10^{-6}

2. 0.00000567, 93200, 0.0001055, 313000

Exercise 4.2

You weigh less at the top of the mountain than at the seashore because you are further from the center of the earth. Hence, the earth exerts less gravitational pull on your body.

Exercise 4.3

The more precise the instrument the more significant figures you can record with the instrument.

Exercise 4.4

The buret will allow you to record more significant digits. The buret is calibrated to the nearest 0.1 ml whereas the graduated cylinder is calibrated to the nearest 1 ml.

Exercise 4.5

23°C or 296°K

Exercise 4.6

Please refer to your chemistry textbook.

Exercise 4.7

36000 calories or 3.6 kilocalories

Exercise 4.8

1. $\dfrac{4 \text{ pints}}{1 \text{ quart}}$ or $\dfrac{1 \text{ quart}}{4 \text{ pints}}$

2. $\dfrac{3600 \text{ sec}}{1 \text{ hour}}$ or $\dfrac{1 \text{ hr}}{3600 \text{ sec}}$

3. $\dfrac{1 \text{ kg}}{1 \times 10^6 \text{ mg}}$ or $\dfrac{1 \times 10^6 \text{ mg}}{1 \text{ kg}}$

Exercise 4.9

Known	→	Connection	→	Goal

$\dfrac{65\ mi}{hr}$ 　　　 $\dfrac{1\ mi}{1.6094\ km}$ or $\dfrac{1.6094\ km}{1\ mi}$ 　　 m/sec

$\dfrac{10^3\ m}{1\ km}$ or $\dfrac{1\ km}{10^3\ m}$

$\dfrac{3600\ sec}{1\ hr}$ or $\dfrac{1\ hr}{3600\ sec}$

$$\dfrac{65\ mi}{hr} \times \dfrac{hr}{3600\ sec} \times \dfrac{1.6094\ km}{1\ mi} \times \dfrac{10^3\ m}{1\ km} = 29\ m/sec$$

Exercise 4.10

3.79×10^{-10}

CHAPTER 5

Learning Inventory 2

1. Elements are homogeneous substances composed of only one kind of atom. Compounds are homogeneous substances composed of two or more elements chemically united in a definite ratio such as H_2O or CH_4. Mixtures are heterogenous substances composed of two or more substances in no definite ratio. The substances that comprise mixtures are not chemically united and may be separated from each other by physical means.

2. An atom is the smallest particle of an element that retains all the properties of that element. Atoms are composed of protons, neutrons, and electrons.

3. Protons are positive particles found in the nucleus of an atom. Neutrons are neutral particles found in the nucleus of an atom. Electrons are negatively charged particles located outside the nucleus in energy shells or levels.

4. The Periodic Table is important to chemists and chemistry students because it contains a great deal of information about the atoms of the elements. You can make predictions about the physical and chemical behavior of the atoms of the elements by knowing their location and position in the table.

5. Atoms and their corresponding ions are similar in that they contain the same number of protons and neutrons. Atoms and their

corresponding ions are different in that the atom is electrically neutral whereas its corresponding ion has either a positive or negative charge.

6. Atoms and ions participate in forming chemical compounds by transferring or sharing valence electrons. Atoms and ions transfer or share electrons in order to achieve a stable octet of electrons, thereby achieving a similar electron arrangement of the noble gas closest to them in the Periodic Table.

Exercise 5.1

1. 13 protons, 14 neutrons, and 13 electrons

2. 35

3. Na

4. Ca

Exercise 5.2

1. the 3 energy level or shell

2. 2 electrons

3. 2,8,7

Exercise 5.3

Lithium	Li^{1+}
Potassium	K^{1+}
Magnesium	Mg^{2+}
Barium	Ba^{2+}
Aluminum	Al^{3+}
Oxygen	O^{2-}
Sulfur	S^{2-}
Fluorine	F^{1-}
Iodine	I^{1-}

Exercise 5.4

Ammonium	NH_4^{1+}
Nitrate	NO_3^{1-}
Nitrite	NO_2^{1-}
Carbonate	CO_2^{2-}
Hydrogen carbonate	HCO_3^{1-}
Phosphate	PO_4^{3-}
Chlorate	ClO_3^{1-}
Sulfate	SO_4^{2-}
sulfite	SO_3^{1-}

Exercise 5.5

Cation	Anion					
	O^{2-}	Cl^{1-}	NO_3^{-1}	SO_4^{2-}	HSO_4^{1-}	PO_4^{3-}
Na^{1+}	Na_2O	$NaCl$	$NaNO_3$	Na_2SO_4	$NaHSO_4$	Na_3PO_4
Mg^{2+}	MgO	$MgCl_2$	$Mg(NO_3)_2$	$MgSO_4$	$Mg(HSO_4)_2$	$Mg_3(PO_4)_2$
Sr^{2+}	SrO	$SrCl_2$	$Sr(NO_3)_2$	$SrSO_4$	$Sr(HSO_4)_2$	$Sr_3(PO_4)_2$
Al^{3+}	Al_2O_3	$AlCl_3$	$Al(NO_3)_3$	$Al_2(SO_4)_3$	$Al(HSO_4)_3$	$AlPO_4$
Fe^{2+}	FeO	$FeCl_2$	$Fe(NO_3)_2$	$FeSO_4$	$Fe(HSO_4)_2$	$Fe_3(PO_4)_2$
Fe^{3+}	Fe_2O_3	$FeCl_3$	$Fe(NO_3)_3$	$Fe_2(SO_4)_3$	$Fe(HSO_4)_3$	$FePO_4$
Zn^{2+}	ZnO	$ZnCl_2$	$Zn(NO_3)_2$	$ZnSO_4$	$Zn(HSO_4)_2$	$Zn_3(PO_4)_2$
Pb^{3+}	Pb_2O_3	$PbCl_3$	$Pb(NO_3)_3$	$Pb_2(SO_4)_3$	$Pb(HSO_4)_3$	$PbPO_4$

CHAPTER 6

Learning Inventory 3

1. Zn and HCl

2. $ZnCl_2$ and H_2

3. Yes. The equation is balanced because there are the same kind and same number of atoms or ions on each side of the arrow. This satisfies the Law of Conservation of Matter.

4. The (s) indicates the reactant Zn is in the solid state. The (aq) indicates the reactant HCl is dissolved in water. The (g) indicates the product H_2 is in the gaseous state.

5. It is needed to balance the equation. It represents 2 moles of HCl reacting.

6. To balance the charge on the compound. In zinc chloride each Zn ion carries a +2 charge and each Cl ion carries a −1 charge. Therefore, 2 Cl ions are needed for each Zn ion.

7. One mole of Zn reacts with two moles of HCl to form one mole of $ZnCl_2$ and one mole of H_2.

8. The HCl would be used up first since it reacts in a ratio of 2 to 1 with Zn.

9. You would need four moles of HCl.

10. $0.25 \text{ moles Zn} \times \dfrac{1 \text{ mole ZnCl}}{1 \text{ mole Zn}_2} = 0.25 \text{ moles of ZnCl}_2$

Exercise 6.1

Both hydrogen and oxygen are diatomic elements and must be written as H_2 and O_2 when uncombined with other elements. The equation is also not balanced.

Exercise 6.2

Answers will vary.

Exercise 6.3

Answers will vary.

Exercise 6.4

1. 120.4
2. 180
3. 84

Exercise 6.5

1. 27 g
2. 11.2 L
3. 9 grams

Exercise 6.6

1. 6.75 moles
2. 110 grams
3. 6.88×10^{-9} grams

Exercise 6.7

1. $Zn + 2HCl \rightarrow ZnCl_2 + H_2$
2. $C_3H_8 + 5O_2 \rightarrow 3CO_2 + 4H_2O$

Exercise 6.8

1. 80 moles
2. 188 moles

Exercise 6.9

1. 276 grams
2. 161 grams

CHAPTER 7

Learning Inventory 4

1. A solute is the chemical substance that is being dissolved; a chemical solvent is the substance doing the dissolving. For example, when salt is placed in water, salt is the solute and water is the solvent. A solute dissolved in a solvent gives a chemical solution.

2. We use the rule that "like dissolves like." Alcohol is a polar solute because it has partial charges on the molecule. Water is a polar solvent because of the O-H bond. Since alcohol and water are both polar molecules they dissolve in each other. We say that water and alcohol are "miscible" in each other. Vegetable oil is a nonpolar substance because there are no partial charges on the molecule. Again, water is a polar molecule and has partial charges. Hence, vegetable oil has no attraction to water and will not dissolve in water. We say that vegetable oil and water are immiscible in each other.

3. A 3% solution of hydrogen peroxide is a solution that contains 3 ml of hydrogen peroxide dissolved in 97 ml of water yielding 100 ml of total hydrogen peroxide solution.

4. A 2.0 M solution of sulfuric acid is a solution that contains exactly 2 moles of sulfuric acid dissolved in 1 liter of water.

5. Percent concentration is defined as grams of solute/100 ml of total solution. We apply the solution pathway:

$$\frac{\text{grams of solute}}{\text{volume of solution}} \times 100 = \% \text{ concentration}$$

$$\frac{12 \text{ g NaCl}}{600 \text{ ml water}} \times 100 = 2\% \text{ solution of NaCl}$$

6. Molarity is defined as moles of solute/liter of solution. We apply the solution pathway:

Change grams of solute to moles/change ml of solution to liters

$$11.5 \text{ g of Al}_2(SO_4)_3 \times \frac{1 \text{ mole Al}_2(SO_4)_3}{315 \text{ g Al}_2(SO_4)_3}$$

$$= 0.0365 \text{ moles of Al}_2(SO_4)_3$$

$$750 \text{ ml of solution} \times \frac{1 \text{ L}}{1000 \text{ ml}} = 0.750 \text{ L solution}$$

$$M = \frac{0.0365 \text{ moles}}{0.750 \text{ L}} = 0.049$$

Exercise 7.1

Answers will vary.

Exercise 7.2

2.7%

Exercise 7.3

3.10 g

Exercise 7.4

643 g

Exercise 7.5

1. Type B molarity problem; 16.9 g KCl.

2. Type C molarity problem; 0.0323 L or 32.3 ml.

3. Type A molarity problem; .360 M.

4. Type D molarity problem; you need 2.5 ml of the concentrated acid to dilute to a total volume of 150 ml.

Exercise 7.6

Molarity = .278; osmolarity = .278 (Note: glucose is a nonelectrolyte and does not ionize in water. You get only one dissolved molecule of glucose per solid glucose molecule.)

CHAPTER 8

Learning Inventory 5

1. Chemical kinetics involves the study of the rates of chemical re-actions and the factors that affect the rate of a chemical reaction.

2. The nature of the reactants, temperature, concentration, and presence of a catalyst.

3. A reversible reaction is a reaction in which the reactants form products and the products break down to reform the reactants. Eventually the forward and backward processes occur at the same rate and the reversible reaction comes to equilibirum.

4. A reversible reaction at equilibrium is one in which the rate at which the reactants from products exactly equals the rate at which the prodcuts decompose to reform the reactants.

5.

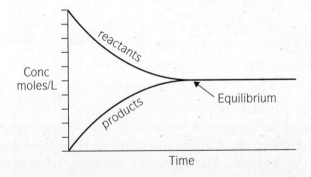

6. When a reversible reaction is in a state of equilibrium, any stress on the system will cause the reaction to shift in a direction to re-lieve the stress. Factors that stress an equilibrium system are changes in concentration, changes in temperature, and changes in pressure.

7.

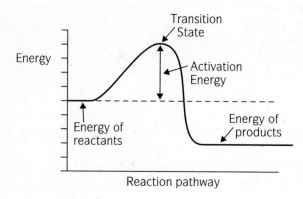

Transition State

Energy

Activation Energy

Energy of reactants

Energy of products

Reaction pathway

8.

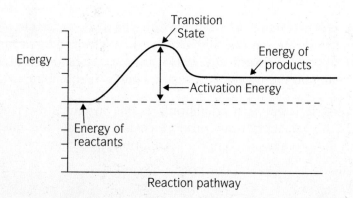

Transition State

Energy

Energy of products

Activation Energy

Energy of reactants

Reaction pathway

9. $K_{eq} = \dfrac{[C]^c \, [D]^d}{[A]^a \, [B]^b}$

Exercise 8.1

1. The energy of the products lies below the energy of the reactants for an exothermic reaction because the products have less energy than the reactants. The difference in energy is equal to the heat energy lost or given up as the exothermic reaction proceeds.

2. The energy of the products lies above the energy of the reactants in an endothermic reaction because the products have more energy than the reactants. The difference in energy is equal to the heat energy gained by the products as the endothermic reaction proceeds.

Exercise 8.2

1. $K_{eq} = \dfrac{[NO_2]^2}{[N_2O_4]}$

2. $K_{eq} = \dfrac{[NO_2]^2 [Cl_2]}{[NO_2Cl]^2}$

Exercise 8.3

$K_{eq} = 0.50$

CHAPTER 9

STATEMENT	A	B
Forms hydronium ion H_3O^+ in water	✓	
Produces solutions with a pH greater than 7		✓
Forms hydroxide (OH^-) ions in water		✓
Is a proton donor	✓	
Is a proton acceptor		✓
Produces solutions with a pH less than 7	✓	
Is an electron pair donor		✓
Is an electron pair acceptor	✓	
Has a $[H_3O^+]$ greater than 1×10^{-7}	✓	
Has a $[OH^-]$ greater than 1×10^{-7}		✓
Reacts with active metals to release hydrogen gas	✓	
Reacts with hydronium ions to produce neutral water		✓
Found in Rolaids®, Tums®, and Alka Seltzer®		✓
Found in gastric juice in the stomach	✓	

1. Acid solutions have a pH less than 7. Basic solutions have a pH greater than 7. Neutral solutions have a pH equal to 7.

2. a salt and water

3. $2HNO_3 (aq) + Ca(OH)_2 (aq) \rightarrow Ca(NO_3)_2 (aq) + 2H_2O$

4.

ACID	K_a VALUE	PK_a VALUE	RATING
H_3PO_4	7.5×10^{-3}	2.1	1
H_3BO_3	7.3×10^{-10}	9.14	3
HCOOH	1.8×10^{-4}	3.75	2

H_3BO_3 is the weakest and H_3PO_4 is the strongest

5. $pH = -\log[H_3O^+]$

$pH = -[\log 3.16 \times 10^{-10}] = 9.5$

Exercise 9.1

ACID	EQUATION ILLUSTRATING IONIZATION OF ACID IN WATER
HCl	$HCl + H_2O \rightarrow H_3O^+ + Cl^-$
HI	$HI + H_2O \rightarrow H_3O^+ + I^-$
HNO_3	$HNO_3 + H_2O \rightarrow H_3O^+ + NO_3^-$
H_2SO_4	$H_2SO_4 + 2H_2O \rightarrow 2H_3O^+ + SO_4^{2-}$
HCOOH	$HCOOH + H_2O \rightleftharpoons H_3O^+ + HCOO^-$

Exercise 9.2

BASE	EQUATION ILLUSTRATING DISSOCIATION OF BASE IN WATER
LiOH	LiHO (s) → Li$^+$ (aq) + OH$^-$ (aq)
NaOH	NaOH (s) → Na$^+$ (aq) + OH$^-$ (aq)
Ca(OH)$_2$	Ca(OH)$_2$ (s) → Ca^{++} (aq) + 2OH$^-$ (aq)
Al(OH)$_3$	Al(OH)$_3$ (s) → Al^{+++} (aq) + 3OH$^-$ (aq)

Exercise 9.3

WEAK ACID	EQUILIBRIUM EXPRESSION FOR DETERMINING K$_a$
HCOOH	$K_a = \dfrac{*[H_3O^+][COO^-]}{[HCOOH]}$
CH$_3$COOH	$K_a = \dfrac{*[H_3O^+][CH_3COO^-]}{[CH_3COOH]}$
HF	$K_a = \dfrac{*[H_3O^+][F^-]}{[HF]}$
C$_6$H$_5$OH	$K_a = \dfrac{*[H_3O^+][C_6H_5O^-]}{[C_6H_5OH]}$

*Note: Your textbook may use [H$^+$] to represent [H$_3$O$^+$]

Exercise 9.4

1. 6.37

2. 4.47×10^{-4}

Exercise 9.5
6.67×10^{-12}

Exercise 9.6

1. 6.3×10^{-3}

2. 5.20

Exercise 9.7

ACID	BASE	EQUATION ILLUSTRATING NEUTRALIZATION
HNO_3	LiOH	$HNO_3\,(aq)+LiOH\,(aq) \rightarrow LiNO_3\,(aq)+H_2O$
HCl	$Ca(OH)_2$	$2HCl\,(aq)+Ca(OH)_2\,(aq) \rightarrow CaCl_2\,(aq)+2H_2O$
H_2SO_4	NaOH	$H_2SO_4\,(aq)+2NaOH\,(aq) \rightarrow Na_2SO_4\,(aq)+2H_2O$
H_3PO_4	KOH	$H_3PO_4\,(aq)+3KOH\,(aq) \rightarrow K_3PO_4\,(aq)+3H_2O$
HCOOH*	NaOH	$HCOOH\,(aq)+NaOH\,(aq) \rightarrow NaHCOO\,(aq)+H_2O$

*The hydrogen comes off because it is attached to an oxygen atom and carries a partially positive charge.

Exercise 9.8
0.182 M acetic acid

INDEX